EBOLA
Clinical Patterns, Public Health Concerns

EBOLA

CLINICAL PATTERNS,
PUBLIC HEALTH CONCERNS

JOSEPH R. MASCI, MD
ELIZABETH BASS, MPh

CRC Press
Taylor & Francis Group
Boca Raton London New York

CRC Press is an imprint of the
Taylor & Francis Group, an **informa** business

CRC Press
Taylor & Francis Group
6000 Broken Sound Parkway NW, Suite 300
Boca Raton, FL 33487-2742

© 2018 by Taylor & Francis Group, LLC
CRC Press is an imprint of Taylor & Francis Group, an Informa business

No claim to original U.S. Government works

Printed on acid-free paper

International Standard Book Number-13: 978-1-4987-1781-6 (Hardback)

Library of Congress Cataloging-in-Publication Data

Names: Masci, Joseph R., author. | Bass, Elizabeth R. (Elizabeth Ruth), 1951- author.
Title: Ebola : clinical patterns, public health concerns / Joseph R. Masci and Elizabeth Bass.
Description: Boca Raton : Taylor & Francis, 2018. | Includes bibliographical references.
Identifiers: LCCN 2017008584 | ISBN 9781498717816 (hardback : alk. paper)
Subjects: | MESH: Hemorrhagic Fever, Ebola--epidemiology | Hemorrhagic Fever, Ebola--prevention & control | Hemorrhagic Fever, Ebola--physiopathology | Ebolavirus
Classification: LCC RC140.5 | NLM WC 534 | DDC 614.5/88--dc23
LC record available at https://lccn.loc.gov/2017008584

Visit the Taylor & Francis Web site at
http://www.taylorandfrancis.com

and the CRC Press Web site at
http://www.crcpress.com

Dedication

The New York City Health and Hospitals Corporation Tiger Team

The Tiger Team was formed to prepare New York City's public hospital system for the possibility of Ebola virus disease (EVD) coming to New York City during the 2014–2016 West Africa outbreak. It was created by Dr. Ross Wilson, medical director of the New York City Health and Hospitals Corporation (NYC H+H) in anticipation of the need for a focused effort combining the skills of a diverse group of NYC H+H professionals. The system that was created permitted extensive training of several thousand personnel culminating in the care of the only Ebola patient to be diagnosed in New York City.

The titles indicated for the members of the Tiger Team are those that they held at the time of the EVD planning efforts described.

Ross Wilson, MD, Medical Director, NYC Health + Hospitals Corporation

Wilson led all aspects of the New York City Health and Hospitals Corporation (NYC H+H) response to Ebola, convening weekly teleconferences on preparedness of all facilities. He chaired weekly or daily conference calls and meetings of the Tiger Team and guided the activities of all its members. He spoke to community groups, gave interviews to the press, and took responsibility for tailoring the NYC H+H to changing circumstances. He assisted in the design of and the development of procedures of the Bellevue unit.

Machelle Allen, MD, Deputy Medical Director, NYC Health + Hospitals Corporation

Allen provided an oversight of overall implementation of the clinical work flows and clinical pathways throughout the enterprise, as well as coordination and collaboration with the New York City Department of Health and Mental Hygiene in ensuring that federal and state guidelines were followed. He evaluated the response to unannounced Ebola exercises at H+H facilities.

Joseph Masci, MD, Director of Medicine, Elmhurst Hospital Center/NYC Health + Hospitals Corporation; Professor of Medicine (Infectious Diseases), Icahn School of Medicine at Mount Sinai

Masci provided guidance on the medical and infection control issues raised by EVD and the types of equipment purchased. He acted as the medical evaluator of unannounced exercises at all of the facilities. He spoke to community groups about EVD.

Lauren Johnston, RN, Chief Nursing Officer, NYC Health + Hospitals Corporation

Johnston expedited the reassignment of staff throughout the system to provide coverage at the Bellevue Hospital Ebola treatment center. She partnered with the nursing union on staffing requirements, evaluating the training, equipment, and supplies of personal protective equipment (PPE). She also provided clinical review for the procurement team on prospective equipment and supplies. In addition, she served as the team leader in the absence of the medical director.

Nicholas Cagliuso, Sr, PhD, MPH, Assistant Vice President, Emergency Management, NYC Health + Hospitals Corporation

Cagliuso oversaw all aspects of communication, coordination, and collaboration with internal and external clinical, operational, and financial partners.

Katie Walker, MBA, RN, Assistant Vice President and Director, Simulation Center, NYC Health + Hospitals Corporation

Walker ensured that PPE training was deployed using the latest evidence-based information from the Centers for Disease Control and Prevention (CDC) statewide sources and local subject-matter experts. She led simulation center educators at the request of the chief medical officer in unannounced incognito embedded patient simulation exercises at each H+H facility to test whether the hospital system was ramped up to correctly assess and either treat or transport as required persons under investigation for EVD.

Roslyn Weinstein, Vice President, Corporate Operations, NYC Health + Hospitals Corporation

Weinstein was responsible for the organization of removal of hazardous waste from the facilities, working closely with the federal government as this involved crossing state lines.

Barbara J. Deiorio, Associate Executive Director, NYC Health + Hospitals Corporation/Jacobi Medical Center

Deiorio developed, directed, and managed all enterprise-wide communications for the Tiger Team during the EVD crisis. This included the creation and development of an internal website, which became the primary vehicle for sharing education, information, updates, and other messaging for 40,000 staff members all across our health system. At its height, the website hosted 6000 unique visitors each day.

Kenra Ford, LSSGB, MBA, Assistant Vice President, Clinical Laboratory Operations, New York City Health + Hospitals Corporation

Ford developed enterprise-wide laboratory guidance documents and supported the development of the laboratory at the Bellevue Hospital Ebola treatment unit.

Paul Albertson, Vice President, Supply Chain Services, NYC Health + Hospitals Corporation

Albertson partnered with the Tiger Team at weekly meetings to standardize the equipment and supply needs, their use within the facilities, and policy and training management, and served as part of the communications team with vendors, external groups, and facility-wide leadership.

Jun Amora, Senior Director of Supply Chain Services, NYC Health and Hospitals Corporation

Amora quickly acquired and managed the EVD personal protective equipment requirements across 11 hospitals and multiple care facilities. As supplies were in short supply across the country, Amora worked with vendors and manufacturers, as well as state and local authorities to manage scarce supplies. He also worked with the Tiger Team to translate rapidly changing CDC guidelines into actual products that met CDC criteria.

Contents

Introduction

Ebola virus disease (EVD) is a rare disorder. It has only been seen in isolated outbreaks since 1976. Until 2014, all cases had been seen in Africa, and all outbreaks had involved fewer, often far fewer, than 500 confirmed cases. Although EVD is highly lethal, with mortality rates exceeding 80% in some outbreaks and settings and averaging 60%, person-to-person transmission of infection is inefficient and appears to require direct contact with body fluids. For these reasons, a global pandemic of EVC seems implausible and spread to developed countries, if it were to occur, would not be expected to result in large-scale outbreaks.

Does an infection seen so infrequently, which seems to pose only a very limited threat of pandemic, warrant a detailed discussion for a broad medical audience as provided in this book? The answer lies in the facts outlined previously. Because EVD is rare, very few health care professionals have been involved in diagnosis or treatment and gained experience in recognizing it. The symptoms and signs of EVD are nonspecific and are common to a large number of infectious diseases that are far more common. Diagnostic tests are not widely available and the approach to diagnosis is familiar to only a few. Further, because person-to-person transmission appears to be by contact with body fluids, rather than airborne, common isolation procedures within hospitals or ambulances, for example, respiratory, airborne, or simple contact precautions, may not be effective in preventing spread to health care workers when the infection is not recognized and extensive barrier precautions are not used. In addition, laboratory and tissue specimens may pose a substantial risk to laboratory personnel prior to confirmation of the diagnosis. As with the case of other rare infections, recognition of EVD may be delayed as other, more likely possibilities are ruled out. For these reasons, EVD represents a potential insidious risk. Clearly, increasing international travel raises the probability that localized outbreaks might cause a multinational or global threat.

Much of the challenge of combating rare contagious diseases lies in formulating adequate strategies to minimize the risk of transmission prior to a confirmed diagnosis. For this reason, the full clinical and epidemiological picture of even the rarest communicable diseases must be taken into account in the formulation of multipurpose, all-hazard containment strategies. The explosive, unexpected nature and high mortality of EVD outbreaks place this infection high on the list of disorders to be considered when such strategies are developed. When other rare infections, such as H5N1 influenza, severe acute respiratory syndrome (SARS), West Nile fever, or the post–9/11 anthrax cases, have been seen, confusion about containment has led to widespread public misperceptions of risk. As seen in the outbreak of 2014–2016, the spread of EVD out of Africa to even a single new country would be expected to result in dire predictions about pandemic spread, travel restrictions, and screening for potential quarantine of those arriving from affected countries. If secondary spread then occurs beyond the initial area impacted, as it did to a very limited extent in the 2014–2016 outbreak, widespread disruption might follow. This effect would be amplified if a health care worker contracted infection, potentially causing other physicians and nurses to fear contact with infected patients. This too was seen in 2014–2016. In addition, as has been seen in other outbreak settings, the *worried well* concerned about their own risk or nonspecific symptoms might flock to emergency departments, thus interfering with normal operations and fueling a sense of panic.

Ready availability of accurate information would help to limit these reactions and inform plans for containment. For these reasons, a thorough and balanced discussion of Ebola virus and the disease it causes is justified. It is hoped that this book will provide that.

The outbreak of EVD that occurred primarily in three countries of West Africa in 2014 through 2016 had an enormous impact on the world in a variety of ways. The fact that the disease had never before been seen in the three countries at the epicenter—Guinea, Liberia, and Sierra Leone—and that health care systems in these countries were extremely limited, caused substantial delays in the local and international response. The number of cases and the duration of the outbreak were both much greater than had been seen previously.

Much was learned about Ebola virus disease during and after the West Africa outbreak of 2014–2016. It is likely that public health authorities have a greater understanding of the demands that an unexpected emergence of a highly dangerous infection imposes on the global health care system.

This book reviews the facts about Ebola virus disease and the key events of the unprecedented 2014–2016 outbreak. Epidemiological, clinical, and laboratory features of the infection are reviewed, as well as the approach to diagnosis.

Recent developments in vaccine strategies are reviewed as well.

The future of EVD is not clear. The fact that an area of the world, West Africa, that had never before documented a case, suddenly became the epicenter of the largest outbreak in history remains incompletely understood.

This book is organized as follows:

Chapter 1: The 2014–2016 Epidemic and Earlier Outbreaks

The history of Ebola in Africa, beginning with the first recognized outbreak in 1976, is reviewed. The emphasis is placed on patterns of spread in the 2014–2016 epidemics.

Chapter 2: Cases outside Africa

Clinical patterns and therapy of patients who contracted Ebola during the West Africa outbreak of 2014–2016 and left, either before or after the diagnosis was confirmed, are summarized.

Chapter 3: Prevention and Containment

Strategies of prevention of transmission of Ebola virus infection as well as containment measures for health care facilities are reviewed.

Chapter 4: Global Response to the Epidemic

The key elements in the regional and international response to the outbreak are reviewed.

Chapter 5: Challenges in the Aftermath of Ebola in West Africa

The medical and social aftermath of the Ebola outbreak in West Africa are reviewed. Topics include restoration of key services as well as long-term medical complications of infection.

Chapter 6: Virus

The structure and replication of the virus as well as the various species that have been identified are reviewed.

Chapter 7: Pathophysiology and Clinical Features of Ebola Virus Infection

The pathophysiology and clinical manifestations of Ebola virus disease are reviewed and characterized.

Chapter 8: Supportive, Antiviral, and Immune Therapy of Patients with Ebola Virus Disease

The key elements in treatment of the patient with Ebola virus disease are reviewed, as well as rehydration and the approach to respiratory and coagulation abnormalities are presented. The potential roles of antiviral and immune therapy are discussed.

Chapter 9: Vaccine Development

Developments in the creation and testing of several vaccines are reviewed.

Chapter 10: Potential Bioterrorism Concerns

The potential role of Ebola virus as an agent of bioterrorism is examined. Topics include plausibility of use and potential means of spread.

Chapter 11: Frequently Asked Questions

Common areas of concern and uncertainty are addressed in a FAQ format. Material from the other chapters of the book is reviewed and summarized.

Chapter 12: Tabletop Exercises for Preparedness

A series of exercises focusing on the aspects of diagnosis and care of patients with Ebola virus infection as well as protection of health care workers is provided.

Appendix

The Appendix provides more detailed information on several issues, including use of personal protective equipment (PPE); assessment of Ebola risk factors; the toll of the West Africa epidemic on health workers; the treatment of Ebola patients outside Africa; the timeline of Ebola outbreaks, and information regarding other special pathogens.

The West Africa Ebola outbreak of 2014–2016 was an enormous human tragedy. Nearly 30,000 cases and more than 11,000 deaths from a completely unexpected infectious disease occurred in a part of the world with very limited health care infrastructure and resources. The acute period of infection—which included disabling fever and diarrhea and, for many patients, life-threatening coagulation disorders and organ system failure—was devastating and, often, not survivable. After the outbreak began, early projections indicated that the number of victims might reach into millions. Health care workers became frequent victims themselves after exposure to patients with advanced infection. Because the outbreak began insidiously and because Ebola remains a relatively rare cause of symptoms that are common to a host of infectious diseases endemic to Africa, there is concern that future outbreaks might not be recognized promptly. The role of international cooperation was critical in bringing the outbreak to a close. As the world turns to other crises, both medical and nonmedical, the lessons of the West African Ebola outbreak of 2014–2016 must be examined and learned.

Authors

Joseph R. Masci, MD, is the director of the Department of Medicine at Elmhurst Hospital Center, a public hospital in New York City, and the professor of medicine (infectious diseases) and of environmental medicine and public health at Icahn School of Medicine at Mount Sinai. He graduated from Cornell University and the New York University School of Medicine. After completing an internship and residency in internal medicine at Boston City Hospital, he completed a fellowship in infectious diseases at the Mount Sinai Medical Center in New York. Since that time, he has been a full-time faculty member at Elmhurst Hospital, the municipal hospital teaching affiliate of the Icahn School of Medicine at Mount Sinai.

During the Ebola epidemic in West Africa in 2014–2016, Dr. Masci served as the infectious diseases physician on the planning committee for Ebola of the New York City Health and Hospitals Corporation (NYC H+H). In this capacity, he helped to design and conduct drills and exercises aimed at preparing the 11 public hospitals in New York City to appropriately diagnose and treat patients with suspected or proven Ebola infection. He also worked with the Tiger Team, a multidisciplinary group created to advise NYC H+H, the largest municipal hospital system in the United States, regarding procedures and equipment to be used in the care of patients with Ebola. In addition, Dr. Masci lectured to a variety of community and professional groups on Ebola.

In other areas of emergency preparedness, Dr. Masci served as chairman of the Emergency Preparedness Council of NYC H+H from the time of the 2001 anthrax attacks through 2009. He coauthored the book *Bioterrorism: A Guide for Hospital Preparedness* (CRC 2005) with Elizabeth Bass and currently acts as the infectious diseases consultant on Zika virus for NYC H+H.

Dr. Masci has conducted two international projects directed at the care of HIV and other infectious diseases in Russia (2004–2011) and in Ethiopia (2008–2014).

He has served as the medical director of AIDS services for the Queens Health Network (1998–2016) and chaired the health workgroup of the New York City Health and Human Services HIV Planning Council from 1998 through 2003. He has published his book *Outpatient Management of HIV Infection* in four editions (Mosby-Yearbook 1992, 1996; CRC Press 2001; Informa Healthcare 2011).

He is the recipient of a number of awards, including the Linda Laubenstein HIV Clinical Excellence Award from the New York State Department of Health AIDS Institute and the Volunteer Service Gold Award from the President's Council on Service and Civic Participation.

Elizabeth Bass, MPh, is a visiting associate professor in the School of Journalism at Stony Brook University, Stony Brook, New York, specializing in teaching science and health journalism and science communication. She was the founding director of the Alan Alda Center for Communicating Science at Stony Brook University, which developed innovative techniques to help scientists and medical professionals around the nation learn to communicate more effectively with people outside their own field. Before coming to Stony Brook University in 2007, she worked as a newspaper editor for almost two decades. For 6 years, she was the science and health editor of *Newsday*, the daily newspaper on Long Island, New York, where she edited coverage of Ebola that won the Pulitzer Prize for explanatory journalism. She also taught journalism as an adjunct at the Columbia University Graduate School of Journalism, New York, and at Hofstra University, East Garden City, New York. Bass was an associate editor of *Human Diseases and Conditions* (Charles Scribner's Sons 2001), a contributor to the *Encyclopedia of Science and Technology*, edited by James Trefil (Routledge 2011), and a coauthor of two books: *KidsHealth Guide for Parents: Pregnancy to Age 5* (Contemporary Books 2002) with Steven A. Dowshen, MD, and Neil Izenberg, MD, and *Bioterrorism: A Guide for Hospital Preparedness* (CRC Press 2004) with Joseph R. Masci, MD. She has a bachelor's degree in English from Cornell University and a master's degree in public health from Stony Brook University.

PART ONE

The History of Ebola

The 2014–2016 Epidemic and Earlier Outbreaks

1

The pattern of Ebola virus disease (EVD) before the 2014–2016 epidemic in West Africa was of sporadic outbreaks in rural areas of East and Central Africa (Feldmann and Geisbert 2011) involving a few dozen to a few hundred cases.

Five species of the Ebola virus have been recognized:

- The *Zaire species*, which was responsible for the West African outbreak of 2014–2016, was first recognized in 1976 and caused most of the past outbreaks of Ebola (Khan 1999).
- The *Sudan species* caused four epidemics in Sudan and Uganda (Onyango et al. 2007; Sanchez et al. 2004).
- The *Bundibugyo species*, first recognized in Uganda in 2007, caused a limited outbreak with a relatively low case-fatality rate (MacNeil et al. 2010; Clark et al. 2015).
- The *Ivory Coast species* has been recognized to cause disease in only one person, who appeared to acquire the infection after performing a necropsy on a chimpanzee found in an area of primate die-off (Formenty et al. 1999).
- The *Reston species* has only been identified in monkeys and pigs in the Philippines. Human infection has been documented by the identification of IgG antibody to the virus in a small number of

individuals with only mild or asymptomatic infection (Miranda et al. 1999). Unlike other strains of the virus, the Reston virus has not been encountered in Africa. From 1989 through 1996, seven small outbreaks of this strain have been recognized in animals (CDC 2016a). These are not included in the data listed in Table 1.1 or shown in Map 1.1.

The source of Ebola outbreaks typically was not established, and transmission often involved health care facilities and workers. The first recognized outbreak of Ebola virus disease (EVD) occurred in Zaire (now Democratic Republic of Congo, DRC) in 1976 (Johnson et al. 1977) and was among the most significant of the Ebola outbreaks until 2014, with more than 300 cases and an observed mortality of 88%. Although the exact origin of the outbreak is not clear, transmission occurred primarily by means of contaminated needles used by health care workers (Ealy and Dehlinger 2016) at a specific hospital. The virus was named after the outbreak that occurred in the area of the Ebola River. Over the next 38 years, 21 outbreaks were recognized; most caused by either the Ebola–Zaire (ZEBOV) or the Ebola–Sudan (SEBOV) strains of the virus (Johnson et al. 1977; WHO 1978a, 1978b; Baron et al. 1983; CDC 2001). All occurred in East Africa, primarily in Democratic Republic of Congo (DRC), Sudan, Gabon, and Uganda, and involved no more than approximately 400 cases. These were typically brought under control in relatively short periods of time, but the largest outbreak prior to 2014, which occurred in Uganda, involved more than 400 cases and lasted for one year (Ealy and Dehlinger 2016). The caseloads and death tolls from these outbreaks are shown in Graphs 1.1 and 1.2.

Despite the relatively small number of cases in these outbreaks, important insight into the epidemiology of EVD was gained. Significantly, it was recognized that close, physical contact was required for transmission and contact within health care facilities as well as burial practices were identified as particularly important factors in spread of the disease. Preliminary work on vaccines was also begun (Bukreyev et al. 2006; Martin et al. 2006; Warfield et al. 2007; Swenson et al. 2008; Geisbert et al. 2009; Tsuda et al. 2011).

Despite the understanding of the Ebola virus and EVD that was gained through investigation of these outbreaks, the West African epidemic of 2014–2016 again raised questions about the means of transmission and necessary steps

TABLE 1.1 Ebola outbreaks in Africa, 1976–2016 (accompanies Map 1.1)

DATE	COUNTRY	TOWN	CASES	DEATHS	SPECIES
June–November 1976	South Sudan	Nzara	284	151	Sudan ebolavirus
September–October 1976	Democratic Republic of the Congo	Yambuku	318	280	Zaire ebolavirus
1977	Democratic Republic of the Congo	Tandala	1	1	Zaire ebolavirus
1979	South Sudan	Nzara	34	22	Sudan ebolavirus
November 1994	Côte d'Ivoire (Ivory Coast)	Tai Forest	1	0	Taï Forest ebolavirus
December 1994–February 1995	Gabon	Mekouka	52	31	Zaire ebolavirus
January–June 1995	Democratic Republic of the Congo	Kikwit	315	250	Zaire ebolavirus
February 1996; July 1996–January 1997	Gabon	Mayibout	37	21	Zaire ebolavirus
1996	Gabon	Booue	60	45	Zaire ebolavirus
November 1996	South Africa	Johannesburg	2	1	Zaire ebolavirus
2000	Uganda	Gulu	425	224	Sudan ebolavirus
October 2001–March 2002	Gabon	Libreville	65	53	Zaire ebolavirus
October 2001–March 2002	Republic of Congo	Unspecified (near Gabon border)	57	43	Zaire ebolavirus
2002	Republic of Congo	Mbomo	143	128	Zaire ebolavirus
2003	Republic of Congo	Mbomo	35	29	Zaire ebolavirus
2004	South Sudan	Yambio	17	7	Zaire ebolavirus

(Continued)

TABLE 1.1 (Continued) Ebola outbreaks in Africa, 1976–2016 (accompanies Map 1.1)

DATE	COUNTRY	TOWN	CASES	DEATHS	SPECIES
May–November 2007	Democratic Republic of the Congo	Luebo	264	187	Zaire ebolavirus
August 2007–February 2008	Uganda	Bundibugyo	149	37	Bundibugyo ebolavirus
December 2008–February 2009	Democratic Republic of the Congo	Luebo	32	15	Zaire ebolavirus
May 2011	Uganda	Luwero District	1	1	Sudan ebolavirus
June–October 2012	Uganda	Kibaale District	11*	4*	Sudan ebolavirus
June–November 2012	Democratic Republic of the Congo	Isiro Health Zone	36*	13*	Bundibugyo ebolavirus
November 2012–February 2013	Uganda	Luwero District	6*	3*	Sudan ebolavirus
August–November 2014	Democratic Republic of the Congo	Multiple	66	49	Zaire ebolavirus
Subtotal			2,411	1,596	
2014–2016	Multiple countries in West Africa	Multiple	28,652	11,325	Zaire ebolavirus
Total			31,063	12,921	

Source: CDC, 2016a, https://www.cdc.gov/vhf/ebola/outbreaks/history/chronology.html, https://www.cdc.gov/vhf/ebola/outbreaks/history/distribution-map.html.

* Numbers reflect laboratory confirmed cases only.

Outside Africa, only three human cases were reported before 2014: One nonfatal case in England in 1976 attributed to a laboratory needlestick injury, and two fatal cases in Russia, one in 1996 and one in 2004, both attributed to laboratory contamination.

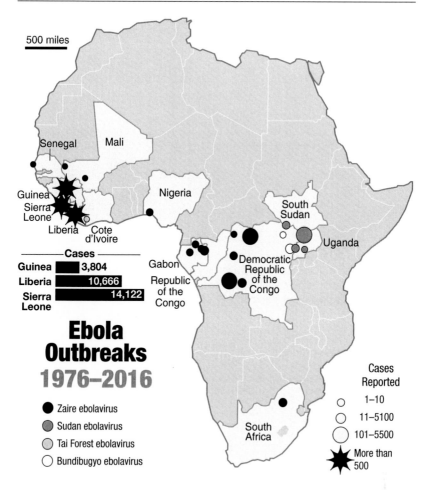

MAP 1.1 Outbreaks since 1976, as listed in Table 1.1. (Map by Rod Eyer.)

for control. Its magnitude and geographical scope were both unprecedented. Graph 1.3 shows how dramatically it outstripped all previous experiences with Ebola. The fact that this outbreak occurred in a region of Africa that had not experienced EVD before, raised concerns that even more widespread disease would occur. The reason for the outbreak of Ebola in West Africa has not yet been identified, although a change in distribution of animal vectors, such as fruit bats, is likely to be a major factor. The magnitude of the outbreak, with more

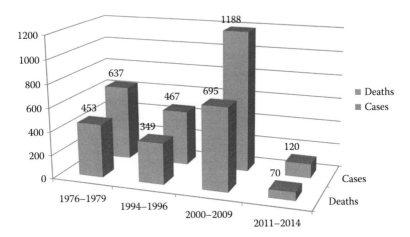

GRAPH 1.1 Ebola Cases and Deaths 1976–2014. Before the West African epidemic of 2014–2016, Ebola had been seen in a total of 24 relatively small outbreaks spread over 38 years in Central or Southern Africa, plus three laboratory cases that occurred in Russia and England.

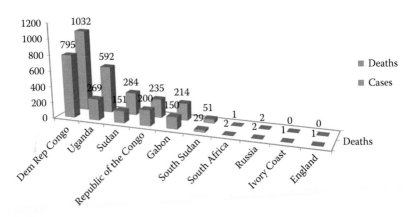

GRAPH 1.2 Ebola Cases and Deaths by Country, 1975–2014. Before the West Africa epidemic of 2014–2016, the burden of Ebola fell mostly on the Democratic Republic of the Congo (Zaire) in Central Africa and Uganda in East Africa.

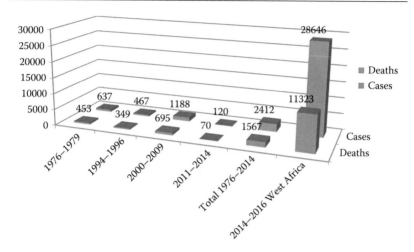

GRAPH 1.3 Ebola Cases and Deaths since 1976. The West Africa epidemic of 2014–2016, which occurred primarily in three countries that had never seen a case before, dwarfed all previous outbreaks.

than 20,000 cases, is also not fully understood, although the fact that the West African cases occurred in or near large cities represented a different pattern than the earlier East African outbreaks.

Although prior Ebola outbreaks had, on occasions, attracted media attention beyond Africa, the magnitude of the 2014–2016 epidemics raised unprecedented concerns all over the world. As discussed in Chapter 3, fears of spread outside West Africa led to measures that restricted travel, screened travelers from the involved countries, and consumed substantial international resources from both governmental and nongovernmental sources. The global reaction also featured high levels of fear and confusion. On a more productive note, research into vaccine development was greatly accelerated and limited data on treatment strategies were developed. On the basis of this experience, it is likely that Ebola will remain a much greater cause for concern among public health officials and governments that it ever was prior to 2014.

OUTBREAKS PRIOR TO 2014

The following information is drawn from CDC accounts (2016b) and from country-specific reports referenced below.

1976

Zaire: Zaire virus. First recognized outbreak. Occurred in Yambuku area. Transmission by close personal contact and contaminated health care needles and syringes.

Cases: 318   Deaths: 280   Mortality rate: 88%

(WHO 1978a)

Sudan: Sudan virus. Occurred in Nzara and Maridi areas. Transmission by close personal contact. Many health care workers infected.

Cases: 284   Deaths: 151   Mortality rate: 53%

(WHO 1978b)

England: Sudan virus. Needlestick injury in laboratory.

Cases: 1   Deaths: 0   Mortality rate: 0%

(Emond et al. 1977)

1977

Zaire: Zaire virus. Case identified retrospectively.

Cases: 1   Deaths: 1   Mortality rate: 100%

(Heymann et al. 1980)

1979

South Sudan: Sudan virus. Same site as the 1976 outbreak.

Cases: 34   Deaths: 22   Mortality rate: 65%

(Baron et al. 1983)

1994

Gabon: Zaire virus. Gold mining camps in the rain forest.

Cases: 52   Deaths: 31   Mortality rate: 60%

(Georges et al. 1999)

Ivory Coast: Scientist infected after performing an autopsy on a chimpanzee in the Tai Forest.

Cases: 1   Deaths: 0   Mortality rate: 0%

(LeGuenno et al. 1995)

1995

Democratic Republic of the Congo: Zaire virus. Outbreak thought to have originated from a worker who had acquired the infection in a forest near Kikwit.

Cases: 315   Deaths: 250   Mortality rate: 81%

1996

Gabon: Zaire virus. Dead chimpanzee was butchered and eaten. Cases occurred among 19 people with direct contact with the animal. The remainder were within their family members.

Cases: 37   Deaths: 21   Mortality rate: 57%

(Georges et al. 1999)

Gabon: Zaire virus. Index case occurred in a hunter possibly acquired from dead chimpanzee.

Cases: 60   Deaths: 45   Mortality rate: 74%

(Georges et al. 1999)

South Africa: Zaire virus. Medical staff member traveled to South Africa after having acquired the infection in Gabon.

Cases: 2   Deaths: 1   Mortality rate: 50%

(WHO 1996)

Russia: Zaire virus. Laboratory contamination.

Cases: 1   Deaths: 1   Mortality rate: 100%

(Borisevich et al. 2006)

2000–2001

Uganda: Sudan virus. Contact through funerals, family members, and patients without appropriate personal protective equipment.

Cases: 425   Deaths: 224   Mortality rate: 53%

(Okware et al. 2002)

2001–2002

Gabon: Zaire virus.

Cases: 65   Deaths: 53   Mortality rate: 82%

(WHO 2003)

Republic of the Congo: Zaire virus. Outbreak on the border of Gabon and the Republic of the Congo.

Cases: 57   Deaths: 43   Mortality rate: 75%

(WHO 2003)

2002–2003

Republic of the Congo: Zaire virus. Outbreak in Mbomo district.

Cases: 143   Deaths: 128   Mortality rate: 89%

(Formenty et al. 2003)

Republic of the Congo: Zaire virus. Outbreak in Mbomo district.

Cases: 35   Deaths: 29   Mortality rate: 83%

(WHO 2004)

2004

South Sudan: Sudan virus. Outbreak in Yambio county of South Sudan.

Cases: 17   Deaths: 7   Mortality rate: 41%

(WHO 2005)

Russia: Zaire virus. Laboratory contamination.

Cases: 1   Deaths: 1   Mortality rate: 100%

(Akinfeyeva et al. 2005)

2007

Democratic Republic of the Congo: Zaire virus. Outbreak in Kasai-Occidental Province.

Cases: 264   Deaths: 187   Mortality rate: 71%

(WHO 2007)

2007–2008

Uganda: Bundibugyo virus. First reported cases of this strain.

Cases: 149   Deaths: 37   Mortality rate: 25%

(MacNeil et al. 2011)

2008–2009

Democratic Republic of the Congo: Zaire virus. Outbreak in Mweka and Luebo health zones of the Kasai-Occidental Province.

Cases: 32   Deaths: 15   Mortality rate: 47%

(WHO 2009)

2011

Uganda: Sudan virus. Outbreak in Luwero District.

Cases: 1   Deaths: 1   Mortality rate: 100%

(Shoemaker et al. 2012)

PHOTO 1.1 During a 2012 Ebola outbreak in Uganda, Red Cross workers put on personal protective equipment (PPE) as they prepare to respond to a report that someone has died of Ebola in a village. (Courtesy of CDC, Atlanta, GA.)

2012

Uganda: Sudan virus. Outbreak in Kibaale District.

Cases: 11   Deaths: 4   Mortality rate: 36.4%

(Albarino et al. 2013)

Democratic Republic of the Congo: Bundibugyo virus.

Cases: 36   Deaths: 13   Mortality rate: 36.1%

(Albarino et al. 2013)

2012–2013

Uganda: Sudan virus. Outbreak in Luwero District.

Cases: 6   Deaths: 3   Mortality rate: 50%

(Albarino et al. 2013)

2014

Democratic Republic of the Congo: Ebola virus. Unrelated to the simultaneous and more extensive outbreak in West Africa.

Cases: 66   Deaths: 49   Mortality rate: 74%

(CDC 2016a)

THE WEST AFRICAN OUTBREAK OF 2014–2016

The West African outbreak was first recognized in the early spring of 2014, and it quickly grew to a magnitude never before seen. The health care systems of the countries most affected were quickly overwhelmed. As a result of steadily increasing and unprecedented levels of aid from international nongovernmental and governmental sources, the epidemic began to recede in the early 2015 and had reached negligible numbers of new cases by a year after it had begun. What follows is a description of how the outbreaks in Guinea, Sierra Leone, and Liberia began and progressed and how health care resources were initially overwhelmed and, ultimately, enhanced.

The initial cases of the outbreak were first recognized in West Africa in March 2014 and occurred in the adjacent countries of Guinea, Liberia, and Sierra Leone almost simultaneously. The Ministry of Health in Guinea reported the first cases from the districts of Guéckédou, Macenta, and Kissidougou and the capital of Conakry on March 21, 2014. By March 30, 2014 the first cases were reported in Liberia from the Foya District. Sierra Leone reported its first cases in May 2014. By June 2014, it was recognized that these three countries were experiencing the worst outbreak of EVD in history (Map 1.2).

Guinea

Guinea was the first country from which cases of EVD were reported in the West African outbreaks of 2014–2016 (Barry et al. 2014; Cadar et al. 2014; Bah et al. 2015). After the outbreak was first recognized in March 2014, Guinea witnessed several surges followed by drops in cases, leading some to think that the epidemic would end quickly. In fact, by mid-April, there was a lull that created hope that the outbreak had been brought under control almost immediately (WHO 2015a). However, global health authorities were skeptical and speculated that the apparent favorable trend represented only the reported cases and did not account for cases in which families

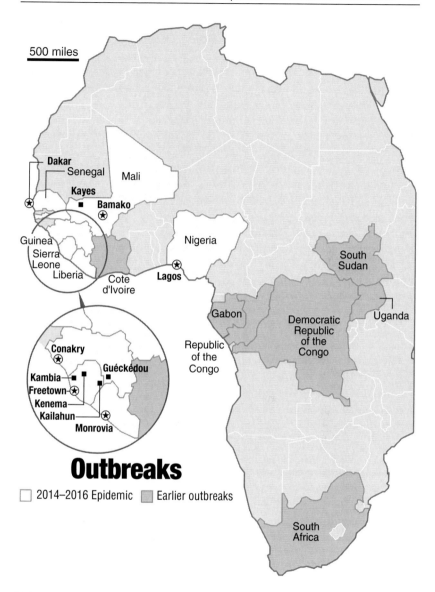

MAP 1.2 Focus of the West African epidemic. (Map by Rod Eyer.)

cared for the sick at home and buried the victims privately and secretly to avoid stigma (WHO 2015a). These families at times refused to allow public health investigators into their homes and communities. By late spring, as treatment centers became more available, recognized cases suddenly rose dramatically, lending some support to the conclusion that the epidemic had been hidden from view rather than truly waning. The need to use indirect data because of logistical and cultural barriers led to inaccurate estimates, likely underestimates, of the extent of the outbreak. Resistance by communities in this fashion occurred in all three of the most impacted West African countries (Thiam et al. 2015; Cohn and Kutalek 2016) but was most pronounced in Guinea. In June 2014, a large resurgence was evident. In addition to resistance to outsiders, violence against health care workers and field response teams became widespread (McCoy 2014). Misinformation and suspicion regarding public health measures represented another barrier to control efforts. The spraying of disinfecting solutions was suspected to be spreading the disease, and concerns were raised that workers from developed countries were causing deaths for sinister reasons. A number of response workers were murdered (WHO 2015a).

Community fears, often based on superstitious beliefs, were exacerbated by inefficiencies of the health care system, resulting in long delays in transporting patients or recovering bodies. Such delays added to the mistrust and strengthened fears that the health care system was responsible for the high death rate. Efforts to identify contacts were suspected of condemning those identified to death (WHO 2015a). A relatively low rate of literacy in Guinea, as well as in Sierra Leone and Liberia, hampered the efforts in public education in the face of a fast-moving crisis.

Critical shortages of supplies, including personal protective equipment, as well as shortages of ambulances and other means of transportation, plagued the response in Guinea. The creation of dedicated treatment centers was significantly delayed, as was the construction of so-called transit centers, where patients under investigation for Ebola could stay while undergoing testing.

As in Sierra Leone and Liberia, the outbreak in Guinea began to come under control in late 2014 and early 2015. The WHO declared Guinea free of Ebola on December 29, 2015, at which time a 90-day period of heightened surveillance was instituted in order to quickly identify any new cases (WHO 2015f). An outbreak of about 10 new cases occurred in March and April, 2016, before Guinea was again declared Ebola-free on June 1, 2016, although the prospect of future resurgences remains (WHO 2016a).

PHOTO 1.2   Treatment center being built in Guinea. (Courtesy of Heidi Soeters, CDC, Atlanta, GA.)

PHOTO 1.3   District health officer, Liberia. (Courtesy of Rebecca Hall, CDC/CDC Connects.)

Liberia

In late March 2014, the first cases of EVD were recognized in Liberia. The outbreak was initially focused in Lofa County and subsequently spread to Montserrado County (CDC 2014a). These areas were the focus of much of the response. An early assessment of preparedness for the treatment of EVD patients was carried out in four adjacent counties that are not yet affected. Each county had a single referral hospital and between 17 and 24 outlying clinics. Prior to the onset of the outbreak, the four counties had a total of six physicians. This fell to three after the first cases were reported. A lack of adequate training was a problem, in addition to absenteeism among nurses and other nonphysician staff, as many did not report for work out of fear of contagion (CDC 2014a). Critical supplies, including sterile and nonsterile gloves, clean water, hand washing stations, soap, bleach, and alcohol-based hand gel, were depleted and communications and transportation between facilities was sparse. Although control efforts had met with some success in the counties first affected (CDC 2014b), the outbreak had reached all the 15 counties of the country by October 2014. As the epidemic expanded throughout the fall, a variety of significant obstacles persisted. These included a lack of trained personnel, including contact tracers, in rural areas as well as continued problems with transportation and communication. Liberia initially was declared to be free of Ebola transmission in January 2016, but experienced a few more cases of Ebola before being declared Ebola-free in June 2016 (WHO 2016b).

Sierra Leone

The outbreak in Sierra Leone began gradually, but peaked quickly in May and June 2015 after several cases were seen among one family returning from Guinea, and surveillance for additional cases was intensified. In June 2015 several cases were traced to the funeral of a traditional healer in the Kailahun district, bordering Guinea. Eventually, more than 300 cases were traced through various chains of transmission to the same funeral (WHO 2015b). A state of emergency was declared in the district and schools and other gathering places were closed. The nearby city of

Kenema soon became a focal point of the outbreak. This progression to a large city has been seen as a key difference from the patterns of spread seen in all the previous outbreaks in East Africa and the reason the West African epidemic of 2014–2016 spread so quickly, became so large, and lasted so long.

Although facilities in Kenema included laboratory and isolation facilities created for Lassa fever, they were quickly taxed beyond their limits with Ebola patients. After a number of health care workers became infected in the district hospital, staffing shortages became critical and were made worse by fears of contagion among medical and nursing providers. Efforts by Médecins Sans Frontières (MSF) (Geneva, Switzerland), also known as Doctors Without Borders, began with the opening of a treatment center in Kailahun, and the World Health Organization established a mobile lab and trained volunteers to aid in contact tracing and communicating with the rural areas. However, because of the rapid acceleration of the outbreak, isolation facilities and laboratories continued to be overtaxed. Quickly deteriorating conditions rendered contact tracing and the identification of chains of transmission difficult, further hampering efforts to get ahead of the expanding crisis. Additional emphasis was placed on the protection and incentivizing of health care workers to maintain staffing at acceptable levels.

As was the case elsewhere in the affected countries, transporting patients from homes to treatment centers was difficult. Because of crowding at those facilities, many were treated supportively at home, at least until test results confirmed the diagnosis of Ebola. This increased the likelihood of further transmission within households. In order to reduce this risk, facilities were created in some areas to allow household members of sick individuals to isolate themselves with others thought to be free of the disease. The World Health Organization, as well as the Red Cross and Red Crescent societies and UNICEF, provided tents (so-called Kenema tents), sleeping mats, and cooking utensils (WHO 2015b). This self-isolation, as it was called, proved to be reasonably effective in preventing further cases.

As the epidemic continued to spread, the first cases were seen in the densely populated capital city of Freetown by late June 2014. Despite the declaration of a national state of emergency, the outbreak spread quickly within the city, reaching 400 newly reported cases per week, as the number of cases began to stabilize in Kenema and Kailahun (Map 1.3).

How the Epidemic Grew

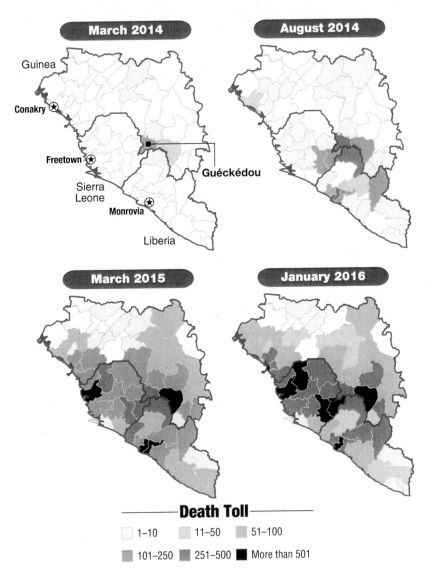

MAP 1.3 How the West African epidemic spread. (Data courtesy of WHO, Geneva, Switzerland. Map by Rod Eyer.)

MALI, SENEGAL, AND NIGERIA: SUCCESS STORIES

In the midst of devastating epidemics of Ebola virus disease in Guinea, Sierra Leone, and Liberia during 2014 and 2015, three neighboring countries, Nigeria, Senegal, and Mali witnessed Ebola outbreaks but were able to contain them before they became widespread. None of these countries had prior experience with EVD. The effective measures of containment employed illustrated several important points. The circumstances and responses in these smaller outbreaks are briefly reviewed here.

Nigeria: In late July 2014, a traveler from Liberia experiencing symptoms of EVD arrived in Lagos by air (WHO 2015c). He was brought to a hospital where he was thought to be suffering from malaria. Nine doctors and nurses contracted EVD and four of them died. The challenge of contact tracing in Lagos, Africa's largest city with a population of 21 million, was great and was further complicated by initial failure to recognize EVD in the traveler. Subsequent to the initial cases, a close contact of the traveler in a second city, Port Harcourt, presented with symptoms. A physician caring for this patient died of Ebola several weeks later. The government acted quickly in collaboration with the World Health Organization to establish isolation facilities in Lagos and Port Harcourt. Initial contact tracing identified hundreds of individuals with possible exposure to EVD. But virtually all contacts were quickly traced and all were determined to be directly or indirectly the result of contact with the initial air traveler from Liberia. Modern virology laboratory facilities allowed for efficient testing, and mass education campaigns were conducted in multiple languages. In the end, these aggressive measures contained the outbreak with a total of 19 cases and 7 deaths, and Nigeria was declared free of Ebola by late October 2014.

Senegal: The first patient with EVD arrived in Dakar, Senegal from Guinea in late August 2014. Dakar is a densely populated city of more than 1 million inhabitants and has sophisticated laboratory facilities, which were approved for testing for hemorrhagic fever viruses. Contact tracing was begun rapidly with assistance from the World Health Organization, the U.S. Centers for Disease Control, and Doctors Without Borders. Contacts were monitored daily and those developing symptoms were tested for Ebola. No transmission occurred, the original patient recovered, and Senegal was declared Ebola-free by late October 2014.

Mali: The first patient with EVD was a two-year-old child from Guinea who was diagnosed in late October 2014. The country had been on alert for the potential arrival of Ebola and had designated a Lassa fever isolation facility in the capital, Bamako, to serve as an Ebola containment unit after six suspected cases, all ultimately found not to have EVD, had been seen several months earlier (WHO 2015c). As in Nigeria and Senegal, modern laboratory facilities were available and aggressive tracing of contacts of the child was conducted. No additional cases were traced to this contact, but in late October 2014 a patient from Guinea with EVD was admitted to a hospital in Bamako. He subsequently died and 433 contacts were identified and evaluated. Seven additional cases, with five deaths, were identified.

Contact tracing in Mali was efficient and effective and was aided by an intensive public education campaign. Mali was declared free of Ebola in January 2015.

The successful containment of Ebola in Nigeria, Senegal, and Mali was the result of several factors. In each country, the infection entered when a single patient arrived from one of the heavily impacted countries in a large city with modern health care and laboratory facilities. This permitted rapid identification of potential contacts and a focused approach to contact tracing and evaluation. In all three countries, this advantage was turned into complete control of the outbreak by aggressive public education and a commitment of government resources. Prior preparation for the arrival of Ebola, particularly in Mali, strengthened the response. As noted above, the medical infrastructure in Guinea, Sierra Leone, and Liberia did not permit early recognition that EVD had arrived, which hampered all efforts at containment.

CASES IN HEALTH CARE WORKERS

A disproportionate share of cases of EVD occurred among health care workers in West Africa (Kilmarx et al. 2014; Matanock et al. 2014; Evans et al. 2015; Fischer et al. 2015; Grinnell et al. 2015; Olu et al. 2015). As shown in Table 1.2, Ebola reduced the health workforce significantly. Factors included the relative lack of training and unavailability of appropriate personal protective equipment. This phenomenon has fueled concerns among health care workers about the risk of contagion in health facilities (CDC 2015). For more on Ebola's toll on African health workers, see the Appendix.

TABLE 1.2 Ebola rates for medical workers versus the general population. Ebola took a disproportionately heavy toll on medical workers during the West African epidemic. Data on medical worker deaths are from May 21, 2015

	GENERAL POPULATION		DOCTORS, NURSES, AND MIDWIVES	
	CASES (% OF POPULATION)	DEATHS (% OF POPULATION)	CASES (% OF WORKFORCE)	DEATHS (% OF WORKFORCE
Guinea	0.03	0.02	2.72	1.45
Liberia	0.25	0.11	10.30	8.07
Sierra Leone	0.21	0.06	10.67	6.85

Source: Evans, D.K. et al., The next wave of deaths from Ebola? The impact of health care worker mortality. World Bank Policy Research Working Paper 7344, July 2015. http://documents.worldbank.org/curated/en/408701468189853698/pdf/WPS7344.pdf.

PHOTO 1.4 A gravedigger at the Bombali Cemetery in Sierra Leone, wearing minimal personal protective equipment in March 2015. By this time, 950 people had been buried in this cemetery, and gravediggers faced intense stigmatization. (Courtesy of Daniel Stowell, CDC/CDC Connects.)

PHOTO 1.5   A burial team from the Health Department in Buchanan, Liberia, heads to a village to perform a safe burial, dressed in full personal protective equipment in April 2015. (Courtesy of Umid Sharapov, CDC/CDC Connects.)

WHY DID THE OUTBREAK OF 2014–2016 BECOME SO EXTENSIVE?

Ebola struck West Africa for the first time in 2014, and the epidemic reached proportions in cases, deaths, and duration that were unprecedented. By the time the outbreak began to wane in early 2015, the impact was 50- to 100-fold greater in the numbers of patients than the largest previous Ebola cluster. The reasons for this are not entirely clear, although this outbreak occurred in circumstances and created conditions that were unique to the countries involved. Among these were the following:

1. The countries affected had not experienced Ebola before. In the countries of East Africa, which had experienced Ebola outbreaks, recognition of an outbreak might have come sooner. Those countries, which had isolation testing and treatment facilities, would likely have been able to mobilize an effective response more efficiently.
2. In all the three countries, Guinea, Sierra Leone, and Liberia, EVD rapidly moved to densely populated urban centers. In fact, the capitals of all the three countries became hotbeds of transmission. This had not

been the case in earlier outbreaks. Although these cities offered the advantage of more modern health care facilities, the number of victims rapidly overwhelmed those resources. Crowded living conditions enhanced transmission and made contact tracing more complicated.

3. Residents of the three countries had a high degree of mobility, including across borders. This tended to enhance spread of EVD as rumors circulated that leaving urban centers for the countryside was protective. This, too, frustrated efforts at contact tracing through which likely exposed persons as well as chains of transmission could be identified and interrupted.

4. It has been pointed out that in previous rapidly accelerating outbreaks, transmission within health care facilities was often the cause (WHO 2015d). In the West African outbreak, community transmission played a proportionately larger role, due in part to burial practices that resulted in close contact with infected corpses and in part to suspicion of outreach health workers. Entire villages were abandoned due to both unchecked transmission of the virus and fears of contagion.

5. The three countries most affected, Guinea, Sierra Leone, and Liberia, are among the world's poorest countries. Table 1.3 presents how they compare in wealth and key health indicators with nations that were less affected by Ebola. The most affected region has recently seen the end of protracted civil wars. The health infrastructure was greatly weakened and largely destroyed, and transportation and communication systems were inadequate to respond to a fast-moving health crisis. These factors also presented extreme barriers to contact tracing, as well as to interrupting chains of transmission by moving individuals to treatment or transit centers. Obviously, the provision of direct care—requiring intravenous fluids and well designed facilities to separate the infected from the uninfected— was also challenged by the infrastructural problems. Sanitation, waste disposal, and provision of protective clothing for health care workers, although crucial in containment strategies, were lacking.

Of course, a profound shortage of health care workers further compounded efforts at control of the epidemic. Before the epidemic, as presented in Table 1.4, Guinea, Sierra Leone, and Guinea already had among the world's lowest per capita rates of doctors and nurses (WHO 2015e). Furthermore, more than 800 health care workers were infected and more than half of them died of EVD. Table 1.5 provides the number of doctors, nurses, midwives, and other health workers who were lost to Ebola during the epidemic. More information on some of the individuals who died can be found in the Appendix of this book.

TABLE 1.3 Health and wealth of Ebola-affected countries

COUNTRY	POPULATION (IN 1,000s)	GROSS NATIONAL PRODUCT PER CAPITA (IN 2015 U.S. DOLLARS [$])	LIFE EXPECTANCY AT BIRTH (YEARS)	MATERNAL MORTALITY (PER 100,000 LIVE BIRTHS)	UNDER-FIVE MORTALITY (PER 1,000 LIVE BIRTHS)	NEONATAL MORTALITY (PER 1,000 LIVE BIRTHS)	NEW HIV INFECTION IN ADULTS 15–49 (PER 1,000 UNINFECTED POPULATION)	TB INCIDENCE (PER 100,000 POPULATION)	MALARIA INCIDENCE (PER 1,000 POPULATION AT RISK)
Heavily Affected Countries									
Guinea	12,609	1,200	51.7	679	93.7	31.3	1.3	177	403.4
Liberia	4,500	900	61.4	725	69.9	24.7	0.6	308	368.8
Sierra Leone	6,453	1,600		1360	120.4	34.9	0.7	310	408.0
Less Affected Countries									
Mali	17,600	2,200	58.2	587	114.7	37.8	1.3	58	460.9
Nigeria	182,202	6,100	54.5	814	108.8	34.3	2.0	322	342.9
Senegal	15,129	2,500	66.7	315	47.2	20.8	<0.1	138	128.1
United States	321,774	55,000	79.3	14	6.5	3.6	—	3.1	—
World	7,313,015	15,700	71.4	216	42.5	19.2	0.5	133	98.6

Sources: World Health Organization, *World Health Statistics, 2016* (population and health statistics); Central Intelligence Agency, *The World Factbook* (GNP figures).

TABLE 1.4   Even before Ebola, there was a lack of medical personnel. Before the 2014–2016 Ebola outbreaks, the three most affected nations already were among the most poorly supplied with doctors, nurses, and midwives of any nations in the world. Dates are the most recent available in the WHO Global Health Observatory data registry

	PHYSICIANS		NURSING AND MIDWIFERY PERSONNEL	
	RANK FROM LOWEST (OUT OF 194)	PHYSICIANS PER 1,000 POPULATION	RANK FROM LOWEST (OUT OF 194)	PERSONNEL PER 1,000 POPULATION
Guinea (2005)	27	0.1	1	0.043
Liberia (2008)	1	0.014	15	0.274
Sierra Leone (2010)	4	0.022	8	0.166

Source: World Health Organization (WHO), Global Health Observatory data repository, density per 1000, data by country, 2015e, http://apps.who.int/gho/data/node.main.

TABLE 1.5   Doctors, nurses, and midwives who died of Ebola. Post-Ebola figures are as of May, 2015. Data on pre-Ebola health workers is from 2004 (nurses and midwives) and 2005 (doctors) for Guinea, 2008 for Liberia, and 2010 for Sierra Leone

	DOCTORS, NURSES, MIDWIVES			
	PRE-EBOLA	POST-EBOLA	% CHANGE	ESTIMATED NUMBERS TO HAVE DIED
Guinea	5395	5317	−1	78
Liberia	1029	946	−8	83
Sierra Leone	1153	1074	−7	79
	7577	7337	−2	240

Source: Evans, D.K. et al., *Lancet*, 3, e439–e440, 2015.

In addition, strikes by health care workers over unpaid wages or concerns about contagion further weakened the response. Traditional healers with unconventional medical practices appeared to be associated with several outbreaks.

6. A variety of cultural beliefs contributed to transmission of the infection. Most often cited are funeral and burial traditions. An analysis by the Ministry of Health, Guinea concluded that 60% of cases in that country could be linked to these practices. In Sierra Leone, estimates were as high as 80% (WHO 2015d). Some extremely

high-risk practices include bathing others with rinse water from corpses or sleeping with the corpse. As reviewed in the sections "The Pathophysiology of Ebola Virus Disease" and "Trends in Viral Studies during the Course of Infection" in Chapter 7, viral concentrations reach their highest level at or near the time of death and the virus remains viable in body fluids after death.

7. An avoidance of hospitals by many, stemming from the high mortality rate associated with EVD and the perception that hospitals were places of death, also permeated some of the cultural responses.

THE END OF THE 2014–2016 EBOLA EPIDEMIC IN AFRICA

On January 16, 2016, after 42 days without a confirmed case, the World Health Organization declared the Ebola outbreak over in West Africa (Kupferschmidt 2016). As of March 27, 2016 the total number of confirmed, probable, and suspected cases in all three of the countries was 28,646. Of these, 10,666 were in Liberia; 14,122 in Sierra Leone; and 3,804 in Guinea. There were a total of 11,323 deaths, for an overall mortality rate of 39.5%, and a mortality rate in confirmed cases of approximately 47% (WHO 2016c). A small number of cases from West Africa were reported in the first few months of 2016, raising concerns that a future surge might occur. Late transmission through sexual contact and prolonged carriage of virus in the semen may prove to be a risk for the ongoing spread.

REFERENCES

Akinfeyeva LA, Aksyonova OI, Vasilyevick IV et al. A case of Ebola hemorrhagic fever. *Infectsionnye Bolezni (Moscow)* 2005;3(1):85–88.

Albarino CG, Shoemaker T, Khristova ML et al. Genomic analysis of filoviruses associated with four viral hemorrhagic fever outbreaks in Uganda and the Democratic Republic of the Congo in 2012. *Virology* 2013;442(2):97–100.

Bah EI, Lamah MC, Fletcher T et al. Clinical presentation of patients with Ebola virus disease in Conakry, Guinea. *N Engl J Med* 2015;372(1):40–47.

Baron RC, McCormick JB, Zubeir OA. Ebola virus disease in southern Sudan: Hospital dissemination and intrafamilial spread. *Bull World Health Organ* 1983;61:997.

Barry M, Traor FA, Sako FB et al. Ebola outbreak in Conakry, Guinea: Epidemiological, clinical and outcome features. *Med Mal Infect* 2014;44(11–12):491–494.

Borisevich IV, Marking VA, Firsova IV et al. Hemorrhagic (Marburg, Ebola, Lassa and Bolivian) fevers: Epidemiology, clinical pictures, and treatment. *Probl Virol (Moscow)* 2006;51(5):8–16.

Bukreyev A, Yang L, Zaki SR et al. A single intranasal inoculation with a paramyxovirus-vectored vaccine protects guinea pigs against a lethal-dose Ebla virus challenge. *J Virol* 2006;80(5):2267–2279.

Cadar D, Gabriel M, Pahlmann M et al. Emergence of Zaire Ebola virus disease in Guinea. *N Engl J Med* 2014;371(15):1418–1425.

CDC. Assessment of Ebola virus disease, health care infrastructure and preparedness—Four counties, Southeastern Liberia, August 2014. *MMWR* 2014a;63(40):891–893.

CDC. Challenges in responding to the Ebola epidemic—Four rural counties, Liberia, August–November 2014. *MMWR* 2014b;63(50):1202–1204.

CDC. Perceptions of the risk for Ebola and health facility use among health workers and pregnant and lactating women—Kenema district, Sierra Leone, September 2014. *MMWR* 2015;63(51 & 52):1226–1227.

Centers for Disease Control and Prevention (CDC). Outbreak of Ebola hemorrhagic fever Uganda, August 2000–January 2001. *MMWR* 2001;50(73):73–77.

Centers for Disease Control and Prevention (CDC). Outbreaks chronology: Ebola virus disease, 2016a. http://www.cdc.gov/vhf/ebola/outbreaks/history/chronology.html.

Centers for Disease Control and Prevention (CDC). Ebola hemorrhagic fever, 2016b. www.cdc.Gov/vhf/ebola.

Clark DV, Kibuuka H, Millard M et al. Long-term sequelae after Ebola virus disease in Bundibugyo, Uganda: A retrospective cohort study. *Lancet Infect Dis* 2015;15(8):905–912.

Cohn S, Kutalek R. Historical parallels, Ebola virus disease and cholera: Understanding community distrust and social violence with epidemics. *PLoS Curr* 2016;8.

Ealy GT, Dehlinger CA. *Ebola: An Emerging Infectious Disease Case Study.* Burlington, MA: Jones & Bartlett Learning, 2016.

Emond RT, Evans B, Bowen ET et al. A case of Ebola virus infection. *BMJ* 1977;2(6086):541–544.

Evans DK, Goldstein M, Popova A. Health-care worker mortality and the legacy of the Ebola epidemic. *Lancet* 2015;3(8):e439–e440. doi:10.1016/S2214-109X(15)00065-0.

Evans DK, Goldstein M, Popova A. The next wave of deaths from Ebola? The impact of health care worker mortality. World Bank Policy Research Working Paper 7344, July 2015. http://documents.worldbank.org/curated/en/408701468189853698/pdf/WPS7344.pdf.

Feldmann H, Geisbert TW. Ebola haemorrhagic fever. *Lancet* 2011;377:849.

Fischer WA, Weber DJ, Wohl DA. Personal protective equipment: Protecting health care providers in an Ebola outbreak. *Clin Ther* 2015;37(11):2402–2410.

Formenty P, Hatz C, Le Guenno B, Stoll A, Rogenmoser P, Widmer A. Human infection due to Ebola virus, subtype Cote d'Ivoire: Clinical and biologic presentation. *J Infect Dis* 1999;179(Suppl 1):S48–S53.

Formenty P, Libama F, Epelboin A et al. Outbreak of Ebola hemorrhagic fever in the Republic of the Congo, 2003: A new strategy? *Med Trop (Marseille)* 2003;63(3):291–295.

Geisbert TW, Geisbert JB, Leung A et al. Single-injection vaccine protects nonhuman primates against infection with Marburg virus and three species of Ebola virus. *J Virol* 2009;83(14):7296–7304.

Georges AJ, Leroy EM, Renaud AA et al. Ebola hemorrhagic fever outbreaks in Gabon, 1994–1997: Epidemiological and health control issues. *J Infect Dis* 1999;179:S65–S75.

Grinnell M, Dixon MG, Patton M et al. Ebola virus disease in health care workers— Guinea, 2014. *MMWR* 2015;64(38):1083–1087.

Heymann DL, Weisfeld JS, Webb PA et al. Ebola hemorrhagic fever. Tandala, Zaire, 1977– 1978. *J Infect Dis* 1980;142(3):372–376.

Johnson KM, Lange JV, Webb PA, Murphy FA. Isolation and partial characterization of a new virus causing acute haemorrhagic fever in Zaire. *Lancet* 1977;1:569.

Khan AS, Tshioko FK, Heymann DL et al. The reemergence of Ebola hemorrhagic fever, Democratic Republic of the Congo, 1995. Commission de Lutte contre les Epidemies a Kikwit. *J Infect Dis* 1999;179(Suppl 1):S76.

Kilmarx PH, Clarke KR, Dietz PM et al. Ebola virus disease in health care workers— Sierra Leone, 2014. *MMWR* 2014;63(49):1168–1171.

Kupferschmidt K. WHO declares Ebola outbreak over, ScienceInsider, January 14, 2016. http://www.sciencemag.org/news/2016/01/who-declares-ebola-outbreak-over.

LeGuenno B, Formenty P, Wyers M et al. Isolation and partial characterization of a new strain of Ebola virus. *Lancet* 1995;345:1271–1274.

MacNeil A, Farnon EC, Wamala J et al. Proportion of deaths and clinical features in Bundibugyo Ebola virus infection, Uganda. *Emerg Infect Dis* 2010;16(12):1969–1972.

MacNeil A, Farnon EC, Morgan OW et al. Filovirus outbreak detection and surveillance: Lessons from Bundibugyo. *J Infect Dis* 2011;204:S761–S767.

Martin JE, Sullivan NJ, Enama ME et al. A DNA vaccine for Ebola virus is safe and immunogenic in a phase I clinical trial. *Clin Vaccine Immunol* 2006;13(11):1267–1277.

Matanock A, Arwady MA, Ayscue P et al. Ebola virus disease cases among health care workers not working in Ebola treatment units—Liberia, June–August 2014. *MMWR* 2014;63(46):1077–1081.

McCoy T. Why the brutal murder of several Ebola workers may hint at more violence to come. *Washington Post*, September 19, 2014. https://www.washingtonpost.com/news/morning-mix/wp/2014/09/19/why-the-brutal-murder-of-eight-ebola-workers-may-hint-at-more-violence-to-come/.

Miranda ME, Ksiazek TG, Retuya TJ et al. Epidemiology of Ebola (subtype Reston) virus in the Philippines, 1996. *J Infect Dis* 1999;179(Suppl 1):S115.

Olu O, Kargbo B, Kamara S et al. Epidemiology of Ebola virus disease transmission among health care workers in Sierra Leone, May to December 2014: A retrospective descriptive study. *BMC Infect Dis* 2015;15:416.

Okware SI, Omaswa FG, Zaramba S et al. An outbreak of Ebola in Uganda. *Trop Med Int Health* 2002;7(12):1068–1075.

Onyango CO, Opoka ML, Ksiazek TG et al. Laboratory diagnosis of Ebola hemorrhagic fever during an outbreak in Yambio, Sudan, 2004. *J Infect Dis* 2007;196(Suppl 2):S193.

Sanchez A, Lukwiya M, Bausch D et al. Analysis of human peripheral blood samples from fatal and nonfatal cases of Ebola (Sudan) hemorrhagic fever: Cellular responses, virus load, and nitric oxide levels. *J Virol* 2004;78:10370.

Shoemaker T, MacNeil A, Balinandi S et al. Reemerging Sudan Ebola virus in Uganda, 2011. *Emerg Infect Dis* 2012;18(9):1480–1483.

Swenson DL, Wang D, Luo M et al. Vaccine to confer to nonhuman primates complete protection against multistrain Ebola and Marburg virus infections. *Clin Vaccine Immunol* 2008;15(3):460–467.

Thiam S, Delamou A, Camara S et al. Challenges in controlling the Ebola outbreak in two prefectures in Guinea: Why did communities continue to resist? *P Afr Med J* 2015;22(Suppl 1):22.

Tsuda Y, Safronetz D, Brown K et al. Protective efficacy of a bivalent recombinant vesicular stomatitis virus vaccine in the Syrian hamster model of lethal Ebola virus infection. *J Infect Dis* 2011;204(Suppl 3):S1090–S1097.

Warfield KL, Swenson DL, Olinger GG, Kalina WV, Amon MJ, Bavari S. Ebola virus-like particle-based vaccine protects nonhuman primates against lethal Ebola virus challenge. *J Infect Dis* 2007;196(Suppl 2):S430–S437.

World Health Organization (WHO). Ebola haemorrhagic fever in Zaire, 1976. *Bull World Health Organ* 1978a;56(2):271–293.

World Health Organization (WHO). Ebola haemorrhagic fever in Sudan, 1976. *Bull World Health Organ* 1978b;56(2):247–270.

World Health Organization (WHO). Ebola haemorrhagic fever—South Africa. *Wkly Epidemiol Rec* 1996;71(47):359.

World Health Organization (WHO). Outbreak(s) of Ebola haemorrhagic fever, Congo and Gabon, October 2001–July 2002. *Wkly Epidemiol Rep* 2003;78(26): 223–225.

World Health Organization (WHO). Ebola hemorrhagic fever in the Republic of the Congo—Update 6. *Wkly Epidemiol Rec* January 6, 2004.

World Health Organization (WHO). Outbreak of Ebola haemorrhagic fever in Yambio, south Sudan, April–June 2004. *Wkly Epidemiol Rec* 2005;80(43):370–375.

World Health Organization (WHO). Ebola virus haemorrhagic fever, Democratic Republic of the Congo—Update. *Wkly Epidemiol Rec* 2007;82(40):345–346.

World Health Organization (WHO). End of the Ebola outbreak in the Democratic Republic of the Congo. Global Alert and Response February 17, 2009.

World Health Organization (WHO). Guinea: The Ebola virus shows its tenacity. January 2015a. http://www.who.int/csr/disease/ebola/one-year-report/guinea/en/.

World Health Organization (WHO). Ebola in Sierra Leone: A slow start to an outbreak that eventually outpaced all others. January 2015b. http://www.who.int/csr/disease/ebola/one-year-report/sierra-leone/en/.

World Health Organization (WHO). Successful Ebola responses in Nigeria, Senegal and Mali. January 2015c. http://www.who.int/csr/disease/ebola/one-year-report/nigeria/en/.

World Health Organization (WHO). Factors that contributed to undetected spread of the Ebola virus and impeded rapid containment. January 2015d. http://www.who.int/csr/disease/ebola/one-year-report/factors/en/.

World Health Organization (WHO). Global Health Observatory data repository, density per 1000, data by country. 2015e. http://apps.who.int/gho/data/node.main.

World Health Organization (WHO). End of Ebola transmission in Guinea. December 29, 2015f. http://www.afro.who.int/en/media-centre/pressreleases/item/8252-end-of-ebola-transmission-in-guinea.html.

World Health Organization (WHO). End of Ebola transmission in Guinea. June 1, 2016a. http://www.afro.who.int/en/media-centre/pressreleases/item/8676-end-of-ebola-transmission-in-guinea.html.

World Health Organization (WHO). End of Ebola transmission in Liberia. June 16, 2016b. http://www.afro.who.int/en/media-centre/pressreleases/item/8699-who-declares-the-end-of-the-most-recent-ebola-virus-disease-outbreak-in-liberia.html.

World Health Organization (WHO). Ebola data and statistics. 2016c. http://apps.who.int/gho/data/view.ebola-sitrep.ebola-summary-20160427?lang=en.

Cases outside Africa

2

With the exception of two nurses in Dallas (Texas) and one in Madrid (Spain) who contracted Ebola virus disease (EVD) after exposure to infected patients from West Africa in 2014, all cases of EVD have resulted from transmission in Africa. During the 2014–2016 West African outbreaks, a small number of health care workers who had acquired the infection in Africa were transferred to treatment facilities in the United States and Europe. Three of fourteen of these patients who were infected in West Africa and transferred to Europe died, as did two of nine people who were infected in West Africa and treated in the United States. The three nurses who contracted EVD outside Africa survived. The potential reasons for this lower mortality rate are discussed in the section "The Lower Mortality Rate of Patients Treated in the United States and Europe" of this chapter. It should be noted that almost all the patients treated in the United States or Europe arrived with the already confirmed diagnosis and all were treated in facilities that could provide intensive care and adequate fluid resuscitation. Although a higher proportion of these patients received antiviral therapy, it is not clear that this played an important role in the higher rates of recovery. It should be noted that a mortality of 22% in these patients was considerable and occurred despite these potential advantages.

UNITED STATES

A total of 11 patients were treated for Ebola infection in the United States during the 2014–2016 West African epidemics. Two were nurses who acquired the infection while caring for a patient in a hospital in Dallas, Texas. These were the only cases of Ebola virus disease ever contracted within the United States. The patient in Dallas and a physician in New York

City, who both had recently traveled from West Africa, were the only other cases of EVD initially diagnosed in the United States. Both of them, however, contracted EVD in West Africa. Of the remaining six patients treated in the United States, five were health care workers and one was a television cameraman brought from West Africa after developing symptoms there. With the exception of the patient treated in Dallas, all patients were treated in specialized facilities designed to minimize the risk of transmission while providing intensive care.

DALLAS

Prior to the fall of 2014, EVD had never been diagnosed or transmitted within the United States. On September 25, 2014, a 45-year-old man, who had arrived from Liberia five days earlier, was presented to an emergency room in Dallas, Texas with fever, abdominal pain, and headache. He was treated for presumed sinus infection and was released (Chevalier et al. 2014). Three days later, he returned to the same emergency room with continued fever and abdominal pain and new onset of diarrhea. He was admitted to the hospital and placed in a private room on standard, contact, and droplet precautions and diagnosed with EVD. Initial investigations, conducted by the Texas Department of State Health Services with assistance from the U.S. Centers for Disease and Prevention (CDC) identified 17 individuals in the community who had close contact either with this patient or with an environment that had potentially been contaminated after the onset of his symptoms. An additional 10 patients were also identified who had been transported in the same ambulance and 21 health care workers who had potential exposure to body fluids without wearing the full personal protective equipment (PPE) that was recommended at the time. Despite supportive therapy, the patient expired on the 11th day after admission. Although questions have been raised about the delay in making a clinical diagnosis of EVD in this case (Fernandez and Bosman 2014; McCann 2014), the facts more likely point to the difficulty in diagnosis of any rare disease. Although hospitals in the United States had been alerted in detail to the possibility of patients with Ebola arriving in their emergency departments, the nonspecific nature of the initial symptoms of EVD make diagnostic errors likely. In addition, the patient had apparently not been aware of his exposure to a person with

EVD in Liberia. Nonetheless, the events in Dallas can be taken to provide support for the need for dedicated special pathogen units where patients with possible exposure histories can be evaluated safely and accurately and cared for effectively.

PATIENTS WHO ACQUIRED EVD IN THE UNITED STATES

Three days later, a 26-year-old nurse, who had participated in the care of the patient hospitalized in Dallas, presented to the hospital's emergency department with fever and was diagnosed with laboratory-confirmed EVD (McCarty et al. 2014). Three days subsequently, a 29-year-old nurse who also had been involved in the first patient's care, presented with fever and rash and was also diagnosed with EVD with laboratory confirmation (McCarty et al. 2014). Because of presumed transmission to the two nurses, several household contacts and a total of 147 health care workers were monitored for 21 days for fever or other potential symptoms of EVD regardless of their PPE use. Eventually, none of the community or hospital contacts of the initial patient or of the two nurses were diagnosed with EVD. Because the second nurse had traveled to Ohio before the diagnosis of EVD was made, a number of individuals potentially exposed during her travel were also monitored. None of these individuals were diagnosed with EVD.

The level of concern regarding the expanding outbreak of EVD in West Africa, which already was growing during the summer of 2014, accelerated dramatically across the United States following the transmission of EVD to the two nurses in Dallas. Initial assessment of the situation by the CDC resulted in confusion regarding the adequacy of PPE worn by those nurses when they cared for the initial patient. Despite assurances that standard contact and droplet precautions were sufficient to prevent transmission of Ebola virus to health care workers, the cases in the Dallas nurses remained were not fully explained. Concerns were also raised about the theoretical possibility of airborne spread of infection. Hospitals and other health care facilities across the country began preparing for the arrival of EVD in a variety of ways, emphasizing early identification of patients arriving from West Africa with fever or other potential symptoms and prompt institution of precautions and appropriate PPE use by staff. CDC guidance regarding appropriate PPE was changed to the recommendation that full skin cover be employed to the staff (Schnirring 2014).

NEW YORK CITY

The last of the four cases ever initially diagnosed in the United States occurred in a physician returning from a humanitarian mission to Guinea. On October 23, 2014, the 33-year-old physician, who was monitoring his temperature at home in New York City since returning from West Africa, noted a fever of 100.3°C and fatigue without other symptoms of EVD (Yacisin et al. 2015; Hartocollis 2015). He indicated that he had used appropriate PPE. He was transported to a specialized unit at Bellevue Hospital Center in Manhattan where he was hospitalized in a previously prepared isolation room and suite. He recovered after several weeks of supportive treatment and antiviral therapy. Community and hospital contacts were monitored and none developed EVD.

EUROPE

Through January 2015, 14 patients with Ebola virus disease (EVD) contracted in West Africa, plus a health care worker who acquired the infection in Spain, were treated in eight countries in Europe. Most were health care relief workers:

Spain: Three (a physician, a missionary, and a nurse); two died
Germany: Three (one physician, one epidemiologist, one lab technician); one died
United Kingdom: Two (both nurses); both recovered
France: Two (one nurse, one health worker); both recovered
Norway: One aid worker; recovered
Switzerland: One doctor; recovered
Italy: One doctor and one nurse; recovered
Netherlands: One member of UN peacekeeping force; recovered

Individual circumstances of the Ebola patients treated outside Africa are outlined on page 257 in the section "Persons Treated for Ebola in the United States and Europe" in the Appendix of this book.

TABLE 2.1 Ebola patients treated outside Africa

COUNTRY WHERE PATIENTS WERE TREATED	TOTAL TREATED	DEATHS	CONTRACTED EVD OUTSIDE AFRICA	PHYSICIANS TREATED	OTHER HCWs TREATED	OTHERS TREATED
United States	11	2	2 Nurses	5	4 (2 Nurses, 1 Hygienist, 1 Clinician)	1 Liberian visiting family in the United States; 1 Journalist
Spain	3	2	1 Nurse	1	1 Nurse	1 Missionary
Germany	3	1	0	1	2 (1 Epidemiologist, 1 Lab technician)	0
France	2	0	0	0	2 (1 Nurse, 1 other HCW)	0
Italy	2	0	0	1	1 Nurse	0
United Kingdom	2	0	0	0	2 Nurses	0
The Netherlands	1	0	0	0	0	1 UN Peacekeeper
Norway	1	0	0	0	0	1 Aid worker
Switzerland	1	0	0	1	0	0
Total	26	5	3	10	10	6

THE LOWER MORTALITY RATE OF PATIENTS TREATED IN THE UNITED STATES AND EUROPE

Five of the 26 patients treated in Europe and the United States died of EVD. This mortality rate of 22% was less than half that was seen in the countries of West Africa, where approximately 47% of patients with confirmed Ebola virus infection died. Although the explanation of this difference has not been established with certainty, a variety of factors are likely to have played a role. Among these unproven possibilities are

1. Health care workers evacuated from West Africa to Europe or the United States may have received care earlier in the course of their infection than individuals living in rural areas in Africa.
2. Use of investigational medications and immune serum was frequent in cases managed in the United States and Europe but essentially nonexistent in cases treated in Africa.
3. Modern laboratory and clinical facilities were not available in Guinea, Liberia, and Sierra Leone. These facilities and capabilities allowed for more rapid confirmation of the diagnosis and monitoring of response to treatment. All but one of the patients treated in the United States were transferred to dedicated isolation and treatment units where intensive care could be provided in a setting to protect staff. The lone patient who was not treated in a dedicated unit died.
4. The availability of intravenous fluids for rehydration, a critical therapeutic strategy, was much greater in the developed countries.
5. It has been suggested that racial disparities, common in many areas of medical access and treatment, played a role in the higher success rate in Ebola cases managed outside of Africa (Kozlowska 2014).

THE SYSTEM OF CARE CREATED IN THE UNITED STATES FOR EBOLA OUTBREAKS

A network of 10 geographically dispersed facilities equipped and staffed to care for patients with Ebola infection and other highly communicable infections was created in the United States in response to the 2014–2016

Designated Treatment Centers
in the UNITED STATES

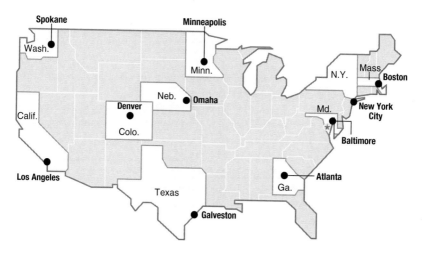

MAP 2.1 In the cities shown here, hospitals and partner health departments were selected in 2015 and 2016 to serve as regional treatment centers for Ebola and other special pathogens. The designated facilities are listed on the next page. (Map by Rod Eyer.)

outbreaks in Africa. Ten hospitals across the United States, as shown in Map 2.1, have been identified and funded for this purpose. Staffing and equipping of these units was aimed at providing protection for hospital physicians, nurses, and other personnel in environments where appropriate care could be safely provided to the patients. In addition, several dozen other hospitals were identified across the United States, were identified that could be staffed and equipped to diagnose and care for patients with Ebola or other highly contagious infections pending the transfer to the higher-level units. The capacity of this system is uncertain. Periodic evaluation of the 10 high-level units is underway, and less frequent evaluation of the other facilities is planned. The types of infections other than Ebola that would be appropriate for this system of care and containment are under discussion and evaluation. Because of the limited capacity of this system of care, with each high-level unit having only a handful of beds, its efficiency would be largely dictated by the number of patients undergoing evaluation and treatment. Efforts to identify, triage, and initiate treatment of Ebola patients in hospitals or other facilities prior to transfer to a specialized hospital unit would be a significant challenge in the setting of a large-scale outbreak.

- *Boston, MA*: Massachusetts Department of Public Health with Massachusetts General Hospital
- *New York, NY*: Department of Health and Mental Hygiene with New York City Health and Hospitals Corporation/H+H Bellevue Hospital Center in New York City
- *Baltimore, MD*: Maryland Department of Health and Mental Hygiene with Johns Hopkins Hospital
- *Atlanta, GA*: Georgia Department of Public Health with Emory University Hospital and Children's Healthcare of Atlanta/Egleston Children's Hospital
- *Minneapolis, MN*: Minnesota Department of Health with the University of Minnesota Medical Center
- *Galveston, TX*: Texas Department of State Health Services with the University of Texas Medical Branch
- *Omaha, NE*: Nebraska Department of Health and Human Services with Nebraska Medicine–Nebraska Medical Center
- *Denver, CO*: Colorado Department of Public Health and Environment with Denver Health Medical Center
- *Spokane, WA*: Washington State Department of Health with Providence Sacred Heart Medical Center and Children's Hospital
- *Los Angeles, CA*: California Department of Public Health with Cedars–Sinai Medical Center

CONCLUSION

Despite fears of a global pandemic, the 2014–2016 Ebola outbreaks remained almost completely confined to West Africa. This appeared to be the result of several key facts. These included the following:

1. Restrictions were placed on travel from the affected countries to the rest of the world after the scale of the outbreak was recognized, as described in Chapter 8. International travel of infected individuals who were not diagnosed before departure, like the Liberian patient who died in Dallas, appeared to have been essentially eliminated.
2. The primary countries involved, Guinea, Liberia, and Sierra Leone, are among the poorest countries in Africa. Travel of individuals from

these countries to other parts of the world is less common than travel between developed countries.

3. Transmission of Ebola virus requires direct physical contact. For this reason, spread within public areas, aside from hospitals and funeral facilities, was not common.

4. The only patients who were evacuated from West Africa for treatment had special status: They were citizens or residents of the United States or Europe, the United Nations workers, or physicians working in Ebola relief. Unlike them, the Liberian patient who died in Dallas came on his own, not knowing he was infected.

It should be recognized that the circumstances of the 2014–2016 West African Ebola outbreak tended to reduce the likelihood of spread to countries outside Africa. In addition, spread to neighboring countries within Africa, including Nigeria, Mali, and Senegal, was limited by effective health care measures (see Chapter 1).

IMPLICATIONS FOR OTHER EMERGING INFECTIONS

The novelty of the Ebola outbreak, its spread to populated centers, and the relative lack of modern medical facilities were all factors in its unprecedented scope and magnitude. These factors may again come together in the future to delay recognition and management of dangerous infectious diseases that are not anticipated. This risk would seem to be greatest in impoverished countries with understaffed health care systems like those initially affected by the Ebola outbreak of 2014–2016. As the world continues to witness the appearance or reemergence of infections, including not only Ebola but also HIV, dengue, Zika, West Nile Virus, and others, the ability of health care systems to recognize and contain novel pathogens will continue to be taxed.

Lessons learned from patterns of emergence, clinical manifestations, and efforts to contain outbreaks caused by these other viral pathogens are instructive in the variety of challenges that they represent.

When compared to these other viral causes of recent international outbreaks, the unique challenge posed by Ebola is clear. It is illustrated by the extraordinarily high mortality of Ebola infection, even when treated in modern facilities, as well as the ease of its transmission to health care workers and household contacts of infected individuals. The need for advanced and elaborate personal protective equipment to protect those caring for its victims is unique in comparison to these other emerging or reemerging infections. Finally, the absence of effective antiviral therapy further serves to underscore the danger of Ebola infection.

What follows is a brief summary of the clinical illnesses, routes of transmission, and implications for the health care system of several other recent emerging or reemerging infections.

Dengue virus, a long-recognized human pathogen, has reemerged as a major health threat in several heavily populated areas of the world. Zika virus was thought to be extremely rare until large outbreaks were recognized in Polynesia and in South America after 2010. West Nile virus arrived unexpectedly in North America in 1999. The patterns of spread of these organisms are particularly instructive. Although all of these viruses are transmitted by mosquitos and, for that reason, carry a greater risk of causing large-scale outbreaks than Ebola virus does, several similarities with the West African Ebola outbreaks of 2014–2016 are worth closer attention.

Dengue Virus

Unlike Ebola virus, Zika virus, and West Nile virus, dengue virus has long been recognized as an important human pathogen. It is currently the most prevalent mosquito-borne viral infection, occurring in many tropical and subtropical regions of the world, and is estimated to cause almost 400 million cases each year, about one-quarter of which are symptomatic. Adults are more likely to develop symptomatic infection than children. Sequential infection with several serologically distinct strains of the virus is much more likely to have severe manifestations, including hemorrhage and shock, than first infections. Dengue infection is transmitted by several mosquito species, the most common of which are *Aedes aegypti,* also the most frequent species to transmit Zika virus, and *Aedes albopictus.*

Zika Virus

Like Ebola virus, Zika virus was recognized as a rare pathogen before a large-scale outbreak occurred in a previously unaffected region of the world, South America, in 2015–2016. Unlike Ebola, it is transmitted by the bite of the mosquito, *Aedes aegypti,* and, rarely, by sexual contact with an infected partner. Most infected individuals have no symptoms and, when symptoms do occur, there are relatively minor and self-limited complaints of arthralgia and myalgia, but when pregnant women are infected, it has been shown to cause devastating fetal abnormalities such as microcephaly. As was the case with the 2014–2015 West African outbreak of Ebola, the 2015 and ongoing outbreak of Zika virus began in a relatively impoverished region. In part because adults, including pregnant women, rarely became symptomatic enough to present for acute medical care, and the effects on the fetus were not initially recognized. Thus, although Zika virus and Ebola virus present very different risks to adults with infection, the involvement of a new geographic area in both the 2014–2016 Ebola outbreak in Africa and the ongoing outbreak of Zika, which originated in Brazil, represented significant challenges to diagnosis and response.

West Nile Virus

West Nile virus, a member of the Japanese encephalitis family that is typically transmitted to humans by the bite of an infected mosquito, first appeared and became endemic in the United States in 1999. Since that time, the virus has been responsible for annual outbreaks of various sizes, totaling approximately 30,000 confirmed cases. A minority of cases show central nervous system involvement, which can include flaccid paralysis or leg weakness as well as rhabdomyolysis, hepatitis, myocarditis, myositis, orchitis, ocular involvement, and multiorgan system failure. Rarely, congenital infection may occur.

REFERENCES

Chevalier MS, Chung W, Smith J et al. Ebola virus disease cluster in the United States—Dallas County, Texas, 2014. *MMWR* 2014;63(46):1087–1088.

Fernandez M, Bosman J. Ebola Victim's Family Blames Hospital and State. *New York Times*, October 11, 2014. https://www.nytimes.com/2014/10/12/us/ebola-victims-family-blames-hospital-and-state.html?_r=0.

Hartocollis A. Doctor who survived Ebola says he was unfairly cast as a hazard and a hero. *New York Times*, February 25, 2015. https://www.nytimes.com/2015/02/26/nyregion/craig-spencer-new-york-ebola-doctor-speaks-out.htm.

Kozlowska H. Has Ebola exposed a strain of racism? *New York Times*, October 21, 2014. https://op-talk.blogs.nytimes.com/2014/10/21/has-ebola-exposed-a-strain-of-racism.

McCann E. Missed Ebola diagnosis leads to debate. *Healthcare IT News*, October 6, 2014. http://www.healthcareitnews.com/news/epic-pushes-back-against-ebola-ehr-blame-shifting.

McCarty CL, Basler C, Karwowski M et al. Response to importation of a case of Ebola virus disease—Ohio, October 2014. Centers for Disease Control and Prevention (CDC). *MMWR* 2014;63(46):1089–1091.

Schnirring L. CDC unveils new PPE guidance for Ebola. CIDRAP (Center for Infectious Disease Research and Policy) News, October 20, 2014. http://www.cidrap.umn.edu/news-perspective/2014/10/cdc-unveils-new-ppe-guidance-ebola.

Yacisin K, Balter S, Fine A et al. Ebola virus disease in a humanitarian aid worker—New York City, October 2014. *MMWR* 2015;64(12):321–323.

Prevention and Containment

3

INTRODUCTION

The outbreak of Ebola virus disease (EVD) that occurred in West Africa in 2014–2016 was, by far, the largest in history. Transmission within homes, health care facilities, and at burial ceremonies added greatly to the number of cases and deaths. Spread of the infection occurred by direct contact with infected individuals and bodies; no convincing evidence of novel forms of transmission was identified. Nonetheless, the frightening rates of severe disease and death and the difficulty of providing effective personal protective equipment to health care workers contributed to exaggerated fears of transmission in the United States.

In addition, anxiety, confusion, and at times, panic arose from misconceptions about how the virus is transmitted. Since Ebola virus disease had been rare before that outbreak, speculation regarding novel routes of transmission and high levels of risk to the general public took root quickly. There had been little suggestion in previous outbreaks that spread could occur by any route other than direct contact with body fluids from a symptomatic patient. Yet concerns that airborne transmission might, nonetheless, be possible (Osterholm et al. 2015) heightened fears of a global pandemic. Similarly, exaggerated concerns that transmission could occur by fomites, such as airline seats, resulted in exaggerated measures to restrict travel of patients and to identify potential exposures among other passengers. Calls for quarantine of returning medical relief workers became a commonplace in the United States (Howell 2014) despite the fact that such measures might discourage the qualified volunteers who were so essential in containing EVD in Africa.

In the countries of Africa that were confronting the epidemic, Sierra Leone, Liberia, and Guinea, fears of spread were, of course, much more legitimate.

Many cases were the result of traditional burial practices that could result in the direct contact with body fluids of the deceased. A study of transmission in Guinea found that in March 2014, 35% (seven of 20) of all transmissions were attributed to hospitals and 15% (three) were attributed to funerals, although these figures soon fell to 9% and 4%, respectively. By comparison, contacts among family members accounted for 72% of transmission (Faye et al. 2015).

Beyond these valid concerns, however, many victims of EVD were inappropriately ostracized and fears that the infection could be contracted by seeking treatment were widespread (Gidda 2014; Karamouzian and Hategekimana 2015; Kobayashi et al. 2015).

Since transmission of Ebola virus does, in fact, require direct contact between body fluid containing the organism in significant quantities and the skin or mucous membranes, containment can be accomplished by adequate barrier precautions and fostered by the separation of infected individuals from other patients and health care facility staff. It is estimated that during September 23–October 31, 2014, interventions such as the creation of Ebola treatment units (ETUs) and community care centers (CCCs), as well as

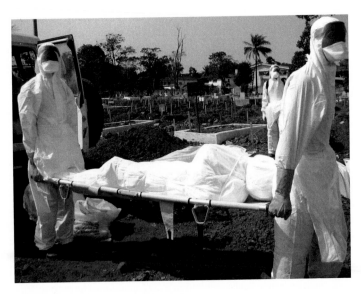

PHOTO 3.1 Shrouded burial. A burial in February 2015, in the King Tom Cemetery in Sierra Leone's capital city of Freetown, shows the kind of procedures that helped stem transmission. The burial team is fully covered with personal protective equipment (PPE), and a burial shroud for a deceased person of the Muslim faith is included. (Courtesy of Carrie Nielsen, CDC/CDC Connects.)

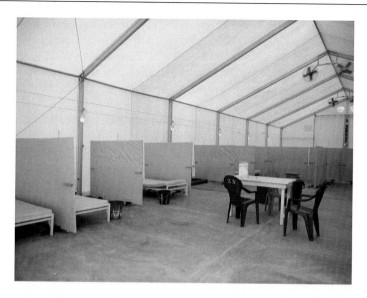

PHOTO 3.2   Ebola unit. Ebola treatment units, like this one in Guinea, shown in 2014, also helped in containing the epidemic. Each patient space includes a low bedframe and two plastic emesis buckets. (Courtesy of Dr. Heidi Soeters, CDC, Atlanta, GA.)

changes in burial techniques, lowered the incidence of EVD in Liberia by approximately 70% compared to what had been projected (Washington and Meltzer 2015).

TRANSMISSION OF EBOLA VIRUS

The essential fact needed to develop strategies to prevent spread of any infectious agent is the means by which it is transmitted. The common ways in which infectious agents are transmitted are as follows (CDC 2016):

Airborne: Inhalation of very small particles in the air containing the agent (e.g., measles, tuberculosis)
Droplet: Inhalation of larger particles (e.g., influenza)
Oral: Placing in the mouth and ingestion (e.g., typhoid fever)

Direct contact: Exposure of unprotected skin or mucous membranes to material containing the infectious agent (e.g., *Herpes simplex* virus infection)

Fomite: A specific form of direct contact in which the contact occurs with an inanimate object in the environment

In a health care setting, the following transmission-based precautions guide the approach to reducing the risk of contagion.

Standard Precautions

Hand washing with alcohol-based hand rub or soap and water before and after each patient contact. This should be conducted even if gloves were worn during contact.

The use of gloves, gowns, surgical masks, goggles, or face shields is dictated by the likelihood of contact with body fluids.

Contact Precautions: To Be Used If Exposure to Body Fluids Is Likely

Standard precautions, plus

- Gloves
- Gown if substantial contact with the patient or their environment is anticipated
- Clean and disinfect room as needed

Droplet Precautions (Respiratory Viruses, *Neisseria meningitidis*, Group A Streptococcus)

Standard precautions, plus

- Place patient in single room with door closed
- Wear face mask such as procedural or surgical mask
- Wear face shield or goggles if aerosolization of secretions is expected
- Provide patient with surgical mask and instruct them to wear it whenever exiting the room or when others are in the room

Airborne Precautions (Tuberculosis, Measles, Chickenpox, *Herpes zoster*)

Standard precautions, plus

- Have patient enter through dedicated isolation entrance if available.
- Place patient in airborne infection isolation room (negative pressure, six air exchanges per hour). If this is not available, instruct patient to wear surgical mask when others are in the room and when the patient exits the room.
- Wear a fit-tested N-95 or high-level disposable respiratory mask.
- Wear gloves, gown, goggles, or face shield as necessary if spraying of respiratory fluids is anticipated.

Transmission of the Ebola virus can occur after apparently minimal but direct exposure to body fluids containing virus. These include blood, feces, sweat, semen and, possibly, saliva (Bausch et al. 2007).

During the course of infection, the concentration of virus in these fluids rises steadily, and postmortem transmission from dead bodies seems to be not only possible but quite common (Gire et al. 2014; Faye et al. 2015). For instance, in Sierra Leone, many cases of EVD occurred among people who attended the funeral of a traditional healer in May 2014 (Gire et al. 2014), and the traditional funeral of a pharmacist in September 2014, appears to have led to a cluster of cases in a previously low-incidence rural region of Sierra Leone (Curran et al. 2016).

For these reasons, the greatest risk of transmission is to close household contacts and sexual partners. The risk to health care workers was significant in the countries of West Africa where the epidemic occurred. In addition, two nurses in Dallas contracted EVD even though they had used personal protective equipment while caring for a patient who became ill after arrival from Liberia and subsequently expired. This raised concern among health care workers even in settings where the potential exposure could be relatively well controlled. The uncertain actual route of transmission of Ebola to these nurses led to extremely detailed and exhaustive strategies to protect all skin and mucous membrane surfaces from coming into contact with any body fluid of patients (CDC 2015a). The U.S. Centers for Disease Control and Prevention developed methods for the safe donning and doffing of such equipment (CDC 2015b), which included repeated training and supervision.

The use of equipment that essentially covered all potentially exposed skin of the health care worker and guidelines for careful removal of this equipment so as to avoid any exposure after patient contact became routine in the United States in cases where Ebola was suspected. No further cases of transmission in the United States were documented.

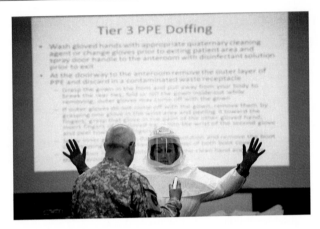

PHOTO 3.3   Personal protective equipment (PPE) training. A training session in how to don and doff PPE, conducted at Tripler Army Medical Center in Hawaii in October 2014. (Courtesy of Staff Sgt. Christopher Hubenthal, American Forces Network, Riverside, CA.)

PHOTO 3.4   Worker drills. Guinean public health care workers take part in an Ebola treatment protocol training drill in 2014. One of the workers crouches to handle a mock Ebola-contaminated towel. Their personal protective equipment includes disposable head covering, transparent face shield, face mask, disposable shirt, apron, gloves, and rubber boots. (Courtesy of Dr. Heidi Soeters, CDC, Atlanta, GA.)

Specific recommendations for PPE for Ebola

As of November 2015, the U.S. CDC provided two categories of recommendations for dealing with persons under investigation (PUIs) for Ebola or with confirmed Ebola (see the Appendix for a detailed description):

Category 1
For patients under investigation who do not have bleeding, vomiting, or diarrhea (CDC 2015a):

1. Single-use (disposable) fluid-resistant gown that extends to at least mid-calf or single-use (disposable) fluid-resistant coveralls without integrated hood.
2. Single-use (disposable) full face shield.
3. Single-use (disposable) face mask.
4. Single-use (disposable) gloves with extended cuffs. Two pairs of gloves should be worn. At least the outer gloves should have extended cuffs.

Category 2
For patients with bleeding, vomiting, diarrhea, or who are clinically unstable or for *any patient with confirmed Ebola* (CDC 2015b), it was recommended that

1. Health care workers caring for patients with Ebola must have received comprehensive training and demonstrated competency in performing Ebola-related infection control practices and procedures.
2. PPE that covers the clothing and skin and completely protects mucous membranes is required when caring for patients with Ebola.
3. Personnel providing care to patients with Ebola must be supervised by an on-site manager at all times, and a trained observer must supervise each step of every PPE donning/doffing procedure to ensure established PPE protocols are completed correctly.
4. Individuals unable or unwilling to adhere to infection control and PPE use procedures should not provide care for patients with Ebola.

This guidance was subsequently updated to

- Expand the rationale for respiratory protection.
- Clarify that the trained observer should not serve as an assistant for doffing PPE.
- Suggest that a designated doffing assistant or *buddy* might be helpful, especially in doffing with the powered air purifying respirator (PAPR) option.

- Modify the PAPR doffing procedure to make the steps clearer.
- Change the order of boot cover removal. Boot covers should now be removed after the gown or coverall.
- Clarify the types of gowns and coveralls that are recommended and provide a link to considerations for gown and coverall selection.
- Emphasize the importance of frequent cleaning of the floor and work surfaces in the doffing area.

PPE SUPPLY AND TRAINING ISSUES

As mentioned above, the complexity of donning and doffing of personal protective equipment for use in the care of patients known or suspected of having Ebola infection is significant. The need to repeatedly train all personnel who might be involved in the care of a patient potentially requires the use of a large amount of this equipment. During the 2014–2016 West African outbreaks, public hospitals in New York City, for example, required frequent retraining of, typically, over 100 staff members in the proper use of the extensive PPE required. The time, equipment, and supervision required for this training was extensive, and the need for simplified approaches to training and more convenient design of equipment was evident.

IMPACT ON HOSPITAL FUNCTIONS

The requirements of effective preparation of a hospital and its staff for the possible arrival of a patient with Ebola infection are significant. A survey of hospital infection control personnel of 357 major U.S. medical centers, conducted during October 2014 by the Society for Healthcare Epidemiology of America (SHEA), revealed that 80% of the time hospital infection prevention and control staff was consumed by Ebola-specific preparations (Morgan 2015) resulting in a 70% reduction in the completion of their tasks. Despite this dedication of services, the ability of hospitals to provide efficient care for patients being evaluated for Ebola infection was limited, as was the care of patients arriving from Africa with other, more common, infections such as malaria and tuberculosis (Parpia et al. 2016).

REFERENCES

Bausch DG, Towner JS, Dowell SF et al. Assessment of the risk of Ebola virus transmission from bodily fluids and fomites. *J Infect Dis* 2007;196(Suppl 2):S142–S147.

CDC. Basic infection control and prevention plan for outpatient oncology settings. 2016. https://www.cdc.gov/hai/settings/outpatient/basic-infection-control-prevention-plan-2011/.

CDC. For U.S. Healthcare Settings: Donning and doffing personal protective equipment (PPE) for evaluating persons under investigation (PUIs) for Ebola who are clinically stable and do not have bleeding, vomiting or diarrhea. November 2015a. https://www.cdc.gov/vhg/ebola/healthcare-us/ppe/guidance-clinically-stable-puis.html.

CDC. Guidance on personal protective equipment (PPE) to be used by healthcare workers during management of patients with confirmed Ebola or persons under investigation (PUIs) for Ebola who are clinically unstable or have bleeding, vomiting, or diarrhea in U.S. hospitals, including procedures for donning and doffing PPE. August 2015b. https://www.cdc.gov/vhf/ebola/healthcare-us/ppe/guidance.html.

Curran KG, Gibson JJ, Marke D et al. Cluster of Ebola virus disease linked to a single funeral—Moyamba District, Sierra Leone, 2014. *MMWR* 2016;65:202–205. doi:10.15585/mmwr.mm6508a2.

Faye O, Boelle P-Y, Heleze E et al. Chains of transmission and control of Ebola virus disease in Conakry, Guinea, in 2014: An observational study. *Lancet Infect Dis* 2015;15(3):320–326. doi:10.1016/S1473-3099(14)71075-8.

Gidda M. Fear and rumors fueling the spread of Ebola. *Time Magazine*, August 12, 2014. http://time.com/3092855/ebola-fear-rumors/.

Gire SK, Goba A, Andersen KG et al. Genomic surveillance elucidates Ebola virus origin and transmission during the 2014 outbreak. *Science* 2014;345(6202):1369–1372. doi:10.1126/science.1259657.

Howell T. Ebola health worker quarantines pit politics against science. *Washington Times*, October 26, 2014. http://www.washingtontimes.com/news/2014/oct/26/ebola-health-worker-quarantines-pit-politics-vs-sc/.

Karamouzian M, Hategekimana C. Ebola treatment and prevention are not the only battles: Understanding Ebola-related fear and stigma. *Int J Health Policy Manag* 2015;4(1):55–56. doi:10.15171/ijhpm.2014.128.

Kobayashi, M, Beer KD, Bjork A et al. Community knowledge, attitudes, and practices regarding Ebola virus disease—Five counties, Liberia, September–October 2014. *MMWR* 2015;64(26):714–718.

Morgan DJ, Braun B, Milstone AM et al. Lessons learned from hospital Ebola preparation. *Infect Control Hosp Epidemiol* 2015;36(6):627–631.

Osterholm MT, Moore KA, Kelley NS et al. Transmission of Ebola viruses: What we know and what we do not know. *mBio* 2015;6(2):e00137-15. doi:10.1128/mBio.00137-15.

Parpia AS, Ndeffo-Mbah ML, Wenzel NS, Galvani AP. Effecs of response to 2014–2015 Ebola outbreak on deaths from malaria, HIV/AIDS, and tuberculosis, West Africa. *Emerg Infect Dis* 2016;22(3):433–441.

Washington ML, Meltzer ML. Effectiveness of Ebola treatment units and community care centers—Liberia, September 23–October 31, 2014. *MMWR* 2015;64(3): 67–69.

Global Response to the Epidemic

4

The Ebola epidemic of 2014–2016 was unique in its magnitude and impact on the world. During the spring and early summer, Ebola virus disease (EVD), which had rightly been regarded as a rare exotic infection, came to spark increasing concern. Countries near to and far from the region directly impacted and reacted to the enlarging outbreak with a mix of reactions ranging from rational concern to fear verging on panic. The high mortality of EVD and early estimates of an epidemic that could result in millions of deaths if left uncontrolled (Meltzer et al. 2014), coupled with unusually relentless media coverage, resulted in extensive efforts to contain the outbreak in West Africa and mitigate its spread there.

The global response to the outbreak of Ebola virus disease (EVD) in West Africa in 2014–2016 occurred on several levels. Nongovernmental aid organizations, in particular Doctors Without Borders (Médecins Sans Frontières, or MSF), were quick to recognize the magnitude and the unique nature of the epidemic and to provide assistance in personnel and equipment to the affected countries. The World Health Organization (WHO) came under criticism for what many considered a delay in acknowledging the extent and the potential global significance of the situation (Westcott 2015). Governmental agencies, including the U.S. Centers for Disease Control and Prevention (CDC) as well as comparable entities from Western Europe, the United Kingdom, and other developed countries, became increasingly involved in the efforts on the ground during the summer and fall of 2014. The U.S. Military developed and maintained a presence and constructed a number of treatment units. In addition to efforts directed at control of the situation in Africa, unfortunately, some elements of the global response contributed to an atmosphere of exaggerated fears of contagion in the developed countries. By fostering extensive travel and quarantine measures, this may have inhibited some efforts to provide resources to Liberia, Guinea, and Sierra Leone in a timely manner and may have unnecessarily discouraged some health care and relief workers from volunteering to go to those countries to help.

What follows is a detailed description of the nature of these and other efforts and a summary of lessons learned from the world's response, which may be applicable to future outbreaks of EVD and other medical emergencies in the developing world.

RESPONSE OF THE WORLD HEALTH ORGANIZATION

Statements by the World Health Organization (WHO) early in the spring of 2014, suggested that the WHO had not fully recognized and acknowledged the rate at which the outbreak was spreading in West Africa and the likelihood that it would be much larger than prior outbreaks of EVD in Africa (Samb 2014; Buchanan 2015). Leaders of the organization subsequently indicated that their initial response had not been based on an accurate assessment of the complexity of the situation in Liberia, Guinea, and Sierra Leone (Boseley 2014). On August 8, 2014, the director general of the WHO, Dr. Margaret Chan, declared the outbreak to be a Public Health Emergency of International Concern (PHEIC) (WHO 2014a). This designation, which had been applied to only two prior outbreaks, the H1N1 influenza epidemic of 2009 and reemergence of polio in 2014, permits the WHO to request member nations of the United Nations (UN) to provide resources for surveillance and control of outbreaks. It indicates that a consensus exists that an international effort is necessary for containment. Under the International Health Regulations (IHR), the WHO must obtain the declaration of a PHEIC from the IHR Emergency Committee.

PHOTO 4.1   Airport screening. A medical worker screens passengers for fever at Sierra Leone's Freetown–Lungi Airport in January 2015. Similar screening was conducted in many airports to identify potentially infected passengers before they could board a plane or enter another country. (Courtesy of Rebecca Myers, CDC/CDC Connect.)

Steps Taken by Individual Countries

International travel restrictions

Several African countries closed their borders to travelers from Guinea, Liberia, and Sierra Leone beginning in August 2014.

This resulted in significant challenges to the transport of needed supplies and personnel. In October 2014, the U.S. Department of Homeland Security began restricting travelers from the affected countries of West Africa to five specific U.S. airports—the international airports in New York, Washington, DC, Atlanta (Georgia), Chicago (Illinois), and Newark (New Jersey)—in order to facilitate screening and tracking of persons arriving from those countries (DHS 2016). The results of this screening are provided in Table 4.1.

TABLE 4.1   Ebola screening at U.S. airports. Travelers who underwent enhanced screening between October 11, 2014 and February 17, 2016. Passengers were referred for screening based on information about where they had traveled

	PASSENGERS REFERRED FOR SCREENING	POSITIVE ANSWER ON CDC QUESTIONNAIRE	ELEVATED TEMPERATURE	TERTIARY SCREENING ("PUBLIC HEALTH ASSESSMENT") BY CDC	TAKEN TO A LOCAL HOSPITAL	FOUND TO BE INFECTED WITH EBOLA
JFK, New York	16,398	582	12	593	10	0
Hartsfield–Jackson, Atlanta, GA	3602	706	9	708	13	0
Washington Dulles, Washington, DC	11,342	1122	47	1139	12	0
Cumulative Total[a]	37,917	2975	81	2996	49	0

Source: U.S. Department of Homeland Security, DHS's coordinated response to Ebola, https://www.dhs.gov/ebola-response, June, 21, 2016.
[a] Includes totals for Chicago O'Hare (IL) and Newark Liberty (NJ) from October 11, 2014 to December 22, 2015.

A variety of airlines, including Air France, British Airways, Emirate Airlines, and Korean Airlines, suspended or restricted flights from the countries involved.

The government of Guinea enacted closures at its borders with Liberia and Sierra Leone with uncertain effect on the outbreak. The effect of border restrictions on the transport of food became significant, and the United Nations World Food Programme of the UN provided assistance in distributing food in the three countries to avert mass starvation (UN World Food Programme 2015). In addition, reductions in foreign investment and trade that resulted from border closing and travel restrictions took a toll. In April 2015, when the epidemic had slowed, the World Bank Group estimated that Guinea, Liberia, and Sierra Leone would, nonetheless, lose at least $2.2 billion in economic growth in 2015 as a result of the epidemic (World Bank 2016).

Quarantine

Quarantine is defined as a process by which individuals exposed to a disease are restricted from travel and separated from the community while they are observed for signs of the disease. In the United States, the federal government may impose quarantine in an effort to contain certain infectious diseases. This strategy was used on a large scale during the 1918 influenza pandemic. It has rarely been used since that time, but hemorrhagic fevers, including EVD, are on a list of so-called quarantinable diseases, which can be a subject to this measure (U.S. Health and Human Services 2009). These include plague, smallpox, cholera, and several others. State laws vary, but states also have power to impose isolation or quarantine, as part of their general police power to preserve public health and safety (CDC 2014).

Quarantine of health care workers returning to the U.S. from West Africa became a controversial issue during a period in the fall of 2014 when unwarranted fear of contagion among the general public reached a peak. Fueled by extensive media coverage and emphasis on the high mortality of EVD, as well as misguided concerns about possible novel forms of transmission (e.g., airborne and fomite), several states ordered such quarantine. In one highly publicized case, a nurse returned to New Jersey after working with EVD victims in Sierra Leone and was placed under mandatory quarantine for three days by state officials despite testing negative for Ebola virus (Miles 2015). One report found that 23 states had quarantine policies that were more stringent than the CDC's guidelines, and that 18 states conducted at least 40 official quarantines, as well as 233 unofficial ones, in which a person went into quarantine or agreed to other restrictions without an official order (ACLU and GHJP 2015).

**BOX 4.1   DISEASES SUBJECT TO
U.S. FEDERAL QUARANTINE**

By Executive Order of the President, federal isolation and quarantine are authorized for these communicable diseases:

- Cholera
- Diphtheria
- Infectious tuberculosis
- Plague
- Smallpox
- Yellow fever
- Viral hemorrhagic fevers (e.g., Ebola)
- Severe acute respiratory syndrome (SARS)
- Flu that can cause a pandemic

Source: U.S. Health and Human Services, What diseases are subject to federal isolation and quarantine law? https://www.hhs.gov/answers/public-health-and-safety/what-diseases-are-subject-to-federal-isolation-and-quarantine, 2009.

Isolation

Isolation is the term applied to the act of separation of an individual known or suspected of having a communicable infection from the community. Isolation to prevent transmission of EVD requires not only separation of the patient from others, but the use of extensive personal protective equipment (PPE) by health care workers and others coming into close contact with an infected individual. During the 2014–2016 EVD outbreaks, isolation was used effectively in West African facilities to prevent transmission. These procedures are based on the knowledge that transmission of Ebola virus requires direct contact with infected body fluid and that transmission by aerosol appears not to occur, and that spread by fomite appears to be unusual at least. Great concern regarding the adequacy of PPE for protection of health care workers arose after two nurses caring for a patient from Liberia in a Dallas hospital contracted the infection, despite apparently wearing some level of protective garb. The reason transmission occurred is not fully understood, although it is thought to have possibly occurred during the removal (doffing) of PPE (Mohan 2014).

CDC guidance for PPE use, as well as recommended procedures for training of health care workers, was revised after these cases.

The special form of PPE recommended to prevent person-to-person transmission of Ebola, as well as the procedures for donning and doffing of those protective garments, is reviewed in Chapter 3 and, in greater detail, in the Appendix.

PHOTO 4.2 Hot lab. Microbiologists from the U.S. Centers for Disease Control and Prevention at work inside the Ebola *hot lab* in Bo, Sierra Leone, in November 2014. (Courtesy of Tara Sealy, CDC/CDC Connects.)

RELIEF AND HEALTH CARE DELIVERY

Many nations responded to the crisis by making financial contributions, as shown in Chart 4.1, and by providing personnel. These included Cuba, which sent several hundred doctors and other health workers, many of whom had experience working in Africa, World Health Organization (2014b) and Uganda, which had several teams who had experience with Ebola in smaller outbreaks. The United States government responded in the spring of 2014 by sending

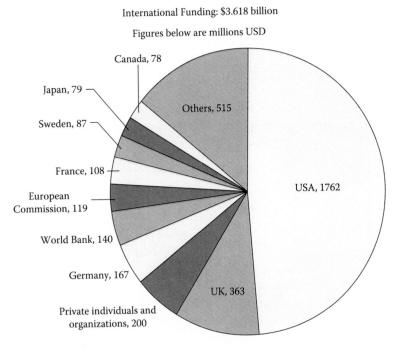

International Funding: $3.618 billion

Figures below are millions USD

Canada, 78
Japan, 79
Sweden, 87
France, 108
European Commission, 119
World Bank, 140
Germany, 167
Private individuals and organizations, 200
Others, 515
USA, 1762
UK, 363

CHART 4.1   International funding to counter the West African Ebola outbreak. The international response included contributions totaling $3.618 billion, as reported by the Financial Tracing Service (FTS), a service managed by the UN Office for the Coordination of Humanitarian Affairs (OCHA). Among the largest of the private donors was Paul G. Allen, philanthropist and cofounder of Microsoft, who gave $58 million and the Bill and Melinda Gates Foundation, which gave $26 million. Data were provided to the FTS by donors or recipient organizations. Figures are in USD millions, rounded to the nearest million. (*Source:* Financial Tracking Service, Ebola Virus Outbreak—West Africa—2014, Table B: Total funding per donor, as of January 7, 2017, http://fts.unocha.org [Table ref: R24]. https://fts.unocha.org/reports/daily/ocha_R24_E16506___1701070230.pdf.)

civilian personnel to the three involved countries to assist in the construction of treatment facilities and burial teams as well as large amounts of supplies, including PPE. In addition, at the height of the epidemic in 2014, 2800 U.S. Military personnel were deployed (Ellis 2015) to further assist in the construction of treatment centers.

RESPONSE OF NONGOVERNMENTAL ORGANIZATIONS

Doctors Without Borders (Médecins Sans Frontières or MSF)

This organization, which already had been working in the region, represented most of the international efforts in the initial months of the outbreak and continues to provide workers, supplies, and coordination of activities in the three affected countries. At its highest level of staffing, MSF had almost 4000 national staff and more than 325 international staff working in the three most affected countries doing contact tracing, health promotion, and surveillance, as well as running Ebola treatment centers. According to the organization, in the first five months of the epidemic, MSF handled more than 85% of all Ebola patients who were hospitalized in the affected countries. Overall, MSF admitted more than 10,000 patients to its Ebola treatment centers, more than half of whom were eventually confirmed as having Ebola (Doctors Without Borders, Undated).

Fast-Tracking of Treatment and Vaccine Trials

Efforts to identify effective treatments and vaccines to combat and prevent EVD accelerated dramatically during the 2014–2016 epidemics. As discussed in Chapter 8, these efforts resulted in preliminary data on several potential antiviral compounds, including the three monoclonal antibody mixture ZMapp. One result of the declining number of cases, beginning in early 2015, was discontinuation of some treatment trials, including that of the antiviral compound brincidofovir. Because of the very limited and largely uninterpretable results of individual instances of the use of several of these agents, the future and best strategy for antiviral therapy remains unclear at this point. However, vaccine trials in various stages have already yielded some clear results. As discussed in Chapter 9, one vaccine in particular, rVSV–ZEBOV appears to confer a high level of resistance to infection and has already entered large-scale trials (Henao-Restrepo 2016).

LESSONS LEARNED FROM THE GLOBAL RESPONSE

The global response to the 2014–2016 West African outbreak of Ebola virus disease underscored several important themes, both positive and negative. On the negative side, it is clear that the international efforts to aid in controlling the epidemic in West Africa were slow in developing. Because of the delayed recognition of the unique nature and magnitude of the outbreak on the part of the WHO, coordinated international efforts did not begin for several months during which the opportunity to prevent widespread transmission was probably missed. This delay calls into question whether the current system for surveilling, monitoring, and responding to emerging infectious diseases is able to provide effective early intervention. The absolutely crucial role of non-governmental organizations (NGOs), particularly Doctors Without Borders, was demonstrated, as it has been in other international health crises. However, the resources and the ability of any NGO to respond on a comparable scale to future crises will always be somewhat unpredictable.

Several additional elements of the global response raised important questions for the future. Although some government responses reflected exaggerated concerns about the risk of contagion of EVD, some incorrectly underestimated this risk. The initial guidance from the U.S. Centers for Disease Control and Prevention (CDC) regarding the necessary procedures to be followed in U.S. hospitals to prevent transmission of Ebola virus infection were not sufficiently detailed. After two nurses at a hospital in Dallas, Texas contracted infection after treating an EVD patient from Liberia, it was clear that general statements regarding droplet and barrier precautions had to be replaced, as they were, by specific guidance on the appropriate type of PPE as well as the proper procedures, including a buddy system for putting the equipment on and for taking it off. Beyond guidance, a great deal of educational information regarding the virus and its modes of transmission, as well as effective screening of individuals presenting to various health care settings for care, had to be provided. A negative lesson should also be learned from the governmental response to fears of contagion and the accentuation of those fears by inconsistent statements and actions by several state and local governments. As demonstrated in prior outbreaks, such as the 2009 H1N1 influenza pandemic, the public takes its lead from media and governmental sources of information. Information from these sources must be meticulously accurate if fear is to be minimized. Even the very limited anthrax attack in the United States shortly after the terrorist attacks of September 11, 2001, resulted in widespread fear and exaggerated portrayals of actual risk. Preventing public concern from reaching inappropriate levels may not be completely possible, but reducing its impact must be a high priority in a crisis environment. If the unlikely had happened and significant numbers of individuals with EVD had presented for care to hospitals in the United States or other developed countries, the impact of the *worried well* might have crippled the emergency response system and fostered broadening of the outbreak in those countries. The public health response was initially overly reassuring concerning the risks to health care workers in direct contact with patients with EVD. The need to retract initial guidance about measures to prevent transmission resulted in a loss of confidence among many. This loss of confidence may have contributed to fears that Ebola could be transmitted in mysterious ways. Airborne transmission was proposed (Osterholm et al. 2015), despite a lack of any supporting evidence. Concerns regarding possible exposure by means of fomites, such as adjacent seats in airplanes became prominent (Engel 2014; Kelly 2014; Murray 2014).

On the positive side, the international efforts, both private and governmental, helped bring the epidemic under control with far fewer than the 500,000–1,000,000 cases that early projections predicted. The combination of efforts to identify contacts of confirmed cases, to protect health care workers, and to reduce transmission through burial practices proved very effective. The logistics of fast tracking of drug and vaccine trials were handled with exceptional efficiency. In the developed countries preparing to receive victims of EVD, specialized containment units were prepared or constructed and procedures for the potential future use of these facilities were brought into focus. Training in the meticulous procedures of donning and doffing PPE when treating patients with EVD could be reinstituted more effectively in future outbreaks based on the experience of 2014–2016. Finally, despite missteps, the international community was effectively brought together, and a coordinated response of relief efforts and quarantine strategies eventually emerged. The singular achievement of developing and piloting an effective vaccine should stand out as an accomplishment that could serve to dramatically reduce the risk of future epidemics of EVD and also to serve as a model of cooperation for future vaccine development.

REFERENCES

American Civil Liberties Union and Yale Global Health Partnership. Fear, politics, and Ebola: How quarantines hurt the fight against Ebola and violate the constitution. December 2015. http://media.wix.com/ugd/148599_3c2a77dfa3c84942be9b40efacda1876.pdf.

Boseley S. World Health Organisation admits botching response to Ebola outbreak. *Guardian*, October 17, 2014. https://www.theguardian.com/world/2014/oct/17/world-health-organisation-botched-ebola-outbreak.

Buchanan E. Ebola crisis: MSF and WHO trade accusations over epidemic response. *International Business Times*, March 25, 2015. http://www.ibtimes.co.uk/ebola-crisis-msf-who-trade-accusations-over-epidemic-response-1493199.

Doctors Without Borders (Medecins sans Frontieres). Ebola Accountability Report 2014–2015. Undated. https://www.doctorswithoutborders.org/sites/usa/files/ebola_accountability_report.pdf.

Ellis R. U.S. bringing home almost all troops sent to Africa in Ebola crisis. *CNN*, February 27, 2015. http://www.cnn.com/2015/02/10/us/ebola-u-s-troops-africa/.

Engel M. Don't fear flying during the Ebola crisis: Experts. *(NY) Daily News*, October 15, 2014. http://www.nydailynews.com/life-style/health/don-fear-flying-ebola-crisis-experts-article-1.1975691.

Henao-Restrepo AM, Camacho A, Longini IM et al. Efficacy and effectiveness of an rVSV-vectored vaccine in preventing Ebola virus disease: Final results from the Guinea ring vaccination, open-label, cluster-randomised trial (Ebola Ca Suffit!). *Lancet* 2016;389:505–518. doi:10.1016/S0140-6736(16)32621-6.

Kelly J. Ebola outbreak: What is risk of catching it on a flight? *BBC News Magazine*, October 16, 2014. http://www.bbc.com/news/world-us-canada-29636355.

Meltzer ML, Atkins CY, Santibanez S et al. Estimating the future number of cases in the Ebola epidemic—Liberia and Sierra Leone, 2014–2015. *MMWR* 2014;63(3):1–14. https://www.cdc.gov/mmwr/preview/mmwrhtml/su6303a1.htm?viewType=Print&viewClass=Print.

Miles SH. Kaci Hickox: Public health and the politics of fear. *Am J Bioethics* 2015;15(4):17–19. doi:10.1080/15265161.2015.1010994.

Mohan G. With Ebola cases, CDC zeros in on lapses in protocol, protective gear. October 15, 2014. http://www.latimes.com/nation/la-na-ebola-protocols-20141015-story.html.

Murray R. Fear among flight attendants after first Ebola diagnosis in U.S. *ABC News*, October 2, 2014. http://abcnews.go.com/Health/fear-flight-attendants-ebola-diagnosis-us/story?id=25919146.

Osterholm MT, Moore KA, Kelley NS et al. Transmission of Ebola viruses: What we know and what we do not know. *mBio* 2015;6(2):00137-15. doi:10.1128/mBio.00137-15.

Samb S. WHO says Guinea Ebola outbreak small as MSF slams international response. Reuters, April 1, 2014. http://www.reuters.com/article/us-guinea-ebola-idUSBREA301x120140401.

United Nations World Food Programme. Ebola response from crisis to recovery. July 2015. http://documents.wfp.org/stellent/groups/public/documents/communications/wfp276313.pdf?_ga=1.17109305.1854183057.1483630612.

U.S. Centers for Disease Control and Prevention. CDC legal authorities for isolation and quarantine. 2014. https://www.cdc.gov/quarantine/aboutlawsregulation squarantineisolation.

U.S. Department of Homeland Security. DHS's coordinated response to Ebola. June 21, 2016. https://www.dhs.gov/ebola-response.

U.S. Health and Human Services. What diseases are subject to federal isolation and quarantine law? 2009. https://www.hhs.gov/answers/public-health-and-safety/what-diseases-are-subject-to-federal-isolation-and-quarantine.

Westcott L. Doctors without borders slams slow international response to Ebola. *Newsweek*, March 23, 2015. http://www.newsweek.com/doctors-without-borders-slams-slow-international-response-ebola-316089.

World Bank. World Bank Group Ebola response fact sheet. April 6, 2016. http://www.worldbank.org/en/topic/health/brief/world-bank-group-ebola-fact-sheet.

World Health Organization (WHO). Statement on the 1st meeting of the IHR Emergency Committee on the 2014 Ebola outbreak in West Africa. August 8, 2014a. http://www.who.int/mediacentre/news/statements/2014/ebola-20140808/en/.

World Health Organization (WHO). WHO welcomes Cuban doctors for Ebola response in West Africa. September 12, 2014b. http://www.who.int/mediacentre/news/statements/2014/cuban-ebola-doctors/en/.

Challenges in the Aftermath of Ebola in West Africa

5

The aftermath of the 2014–2016 Ebola epidemics in West Africa presented a number of significant challenges to the governments of the region as well as to the design of efforts by the global health community. Recovery from this epidemic itself must include measures to reduce the risk of the reemergence of this catastrophic disease. This must include more effective means of surveillance, identification of chains of transmission, and contact tracing. Systems of care must be greatly improved in order to provide facilities that can serve the standard and critical care needs of patients suffering from Ebola virus disease (EVD) while protecting caregivers, and progress toward effective antiviral therapy that must be stimulated. Strategies to best utilize the knowledge that has been gained thus far from vaccine trials must also be developed. Issues surrounding the potential for long-term infection, including persistent sexual transmission or sustained or recrudescent infection in *protected* sites in the body, such as the eye and the central nervous system, will have to be confronted, both epidemiologically and clinically.

In addition, systems of care that were severely impacted by the Ebola epidemic must be rebuilt and strengthened beyond their preepidemic level. HIV/AIDS, malaria, and tuberculosis (TB) care, for example, suffered greatly during the epidemic. The consequences of a failure to continue progress against these diseases, and to risk a large-scale expansion of their reach, risks costing many more lives than the Ebola epidemic did.

Finally, the health infrastructure in the impoverished countries of Guinea, Sierra Leone, and Liberia must be strengthened to advance toward the modern medical world of strategies to reduce the incidence of cancer, cardiovascular disease, and diabetes as well as to improve children's health and the state of obstetrical care. The difficulty of this task was heightened by the economic toll taken by the epidemic (Table 5.1).

TABLE 5.1   The economic toll of Ebola in primary countries. The World Bank estimated the amount of gross domestic product (GDP) lost because of Ebola in the three most hard-hit countries. The medium-term estimate shown here is based on a *low Ebola* scenario in which the caseload reached around 20,000, and the disease was largely contained during the first quarter of 2015. This scenario is fairly close to what occurred, although the caseload reached more than 28,000

	PROJECTED GDP 2014 (IF NO EBOLA)	SHORT-TERM LOSS (2014), IN DOLLARS	MEDIUM-TERM LOSS (2015), IN DOLLARS	CUMULATIVE IMPACT, 2014 AND 2015, IN DOLLARS
Guinea	6.471 billion	130 million (2.1%)	43 million (0.7%)	173 million
Liberia	2.066 billion	66 million (3.4%)	115 million (5.8%)	181 million
Sierra Leone	5.486 billion	163 million (3.3%)	59 million (1.2%)	222 million
Total of three countries	14.023 billion	359 million	129 million	488 million

Source: World Bank, *The Economic Impact of the 2014 Ebola Epidemic: Short- and Medium-Term Estimates for West Africa*, World Bank, Washington, DC, 2014. License: Creative Commons Attribution CC BY 3.0 IGO.

All of these missions represent daunting challenges that cannot be effectively met in a short period of time, but the legacy of the Ebola epidemic of 2014–2016 could be, in part, progress toward these goals. What follows is an overview of some of these challenges and the strategies in development to meet them.

MITIGATING THE THREAT OF A RETURN OF EBOLA

The efforts taken and strategies created in confronting the West African Ebola outbreak of 2014–2016, once largely mobilized, reduced the incidence of new cases from more than 950 cases per week to a case every few months (WHO 2015).

The larger question of how to predict and mitigate future outbreaks of Ebola and other infectious diseases, however, has not yet been answered. It has been pointed out that most pandemics, as with the case of Ebola, originate in animals and are caused by viruses (Morse et al. 2012), and that such outbreaks appear to be increasing with time (Jones et al. 2008). Pandemic influenza, severe acute respiratory syndrome (SARS), West Nile fever, and HIV/AIDS all have followed this pattern and none were predicted to cause human outbreaks before they did. These outbreaks have often been attributed to environmental or behavioral changes among the animal hosts and/or the human population at risk.

The international efforts to bring the West African Ebola epidemic to an end were extensive and costly. It has been pointed out, however, that efforts to control outbreaks typically are costlier than measures to prevent them (Castillo-Chavez et al. 2015).

OTHER MEDICAL CONSEQUENCES OF EBOLA VIRUS INFECTION

The magnitude of the West African Ebola outbreak has resulted in the largest number of survivors of acute infection with this virus ever seen. For this reason, the late manifestation of infection, both known and, perhaps, not previously known, may become common issues in the involved countries of West Africa.

Eye Infection

Eye involvement has been seen frequently in Ebola infection. Symptoms that may be seen during acute infection include conjunctivitis and subconjunctival hemorrhage (Bwaka et al. 1999). During the convalescent phase of infection, uveitis had been described prior to the West African epidemic of 2014–2016 (Kibadi et al. 1999). Necrotizing scleritis and conjunctivitis as well as pathologic involvement of the brain has been described in an Ebola-infected rhesus macaque (Alves et al. 2016). The incidence of an eye involvement in the West African epidemic is not clear, although persistence of Ebola virus in ocular fluid during convalescence nine weeks after resolution of viremia has been documented (Varkey et al. 2015).

Sexual Transmission

The Ebola virus has been detected in the semen of survivors (Bausch et al. 2007) after recovery and sexual transmission has been strongly implicated (Mate et al. 2015). On the basis of studies prior to the West African epidemic, it was thought that this risk could be avoided by sexual abstinence or use of condom for 90 days after infection (Soka et al. 2016). However, Ebola virus RNA similar to that which was isolated from a woman who died of EVD was identified in a male sexual partner 199 days after he had recovered from EVD

(Christie et al. 2015). In order to determine the duration of seminal carriage of Ebola virus in men recovering after the 2014–2016 West African epidemic, Soka and colleagues (Soka et al. 2016) enrolled 466 survivors from several sites in Liberia. Of the 429 subjects who could be fully evaluated, 38 (9%) had one or more semen samples test positive for viral RNA, including 24 (6%) tested positive more than 12 months after the acute infection. The longest time interval between acute infection and detection of persistent Ebola virus in the semen was 565 days. Men who were found to be chronic carriers of Ebola virus did not differ from the others in terms of symptoms of sexually transmitted disease. The median age for men with positive results was 40, compared to 32 among those who did not have a positive test of semen. It was not determined if other health factors, such as HIV infection, affected the risk of prolonged seminal carriage of Ebola virus. On the basis of these results the authors recommended that care for survivors should include semen testing and behavioral counseling.

Central Nervous System Infection

Central nervous system manifestations of Ebola virus infection are common. These may include headache, confusion, coma (Chertow et al. 2014), and seizures, although the role of direct infection of the brain in these manifestations is not clear because evaluation of cerebrospinal fluid (CSF) has rarely been conducted. However, Ebola virus was found in the CSF of a 21-year-old man with stupor, neck stiffness, and seizures who underwent lumbar puncture to evaluate for bacterial meningitis (Sagui et al. 2015).

RESTORING CRITICAL SERVICES

The West African Ebola epidemic had a profound impact on the delivery of services for three key infectious diseases: HIV/AIDS, tuberculosis, and malaria.

Misconceptions about the risk of acquiring Ebola in health care facilities impacted the use of these facilities for prenatal and postnatal care, immunizations, treatment of malaria, and other conditions (Ansumana et al. 2016). Greatly compounding this was the direct impact of Ebola on the health care workforce. Many physicians, nurses, and ancillary staff were infected with Ebola and many of them died. Between January 2014 and March 2015, 815 estimated and probable cases occurred among health care workers in the three countries, according to the WHO, including 328 in Sierra Leone, 288 in Liberia, and 199 in Guinea (Ansumana et al. 2016).

HIV/AIDS

The outbreak of EVD has had a devastating effect on the health care systems in Guinea, Sierra Leone, and Liberia. HIV care, which had already been hampered by rising levels of resistance to antiretroviral therapy (ART) (Loubet et al. 2014) in West Africa as well as significant obstacles to both treatment and prevention (Bhoobun et al. 2014) prior to the arrival of EVD, in particular, was impacted severely.

By June 2014, the three largest hospitals in Monrovia (Redemption, Kennedy, and Saint Joseph) were forced to reduce most clinical activities in order to protect workers from EVD (Tattevin et al. 2015). These conditions persisted for at least eight months after the beginning of the Ebola outbreak, despite the decreasing numbers of cases. Loss of key hospital leaders, as well as strikes and refusal to work by employees pending the provision of adequate personal protective equipment (PPE), are thought to have increased mortality from other diseases. Heroic efforts to maintain HIV patients on ART were made by the staff of the HIV treatment clinics within these hospitals, and administrative offices were turned into clinics in some areas. However, because of precautions instituted around phlebotomy in order to reduce the likelihood of transmission of Ebola virus to phlebotomists, counseling and testing initiatives as well as laboratory monitoring of persons with HIV were negatively impacted. These experiences in Monrovia were mirrored in the rest of Liberia. Fear of EVD among health care workers and the reallocation of resources from HIV to EVD resulted in the closing of 60% of 144 HIV care centers in Liberia by November 2014 (IRIN 2014). It has been estimated that there was an increase of 9%–16% of HIV/AIDS related deaths as a result of the Ebola epidemic in West Africa (Parpia et al. 2016).

HIV in Liberia and Sierra Leone prior to 2014

In 2013, the most recent year for which complete data are available, approximately 30,000 adults and children in Liberia and 57,000 in Sierra Leone were living with HIV (WHO 2014). Deaths due to AIDS were 2700 and 3100, respectively and approximately combined 60,000 AIDS orphans were living in the two countries. In Liberia, approximately 70% of patients were undergoing treatment.

The provision of HIV care in West Africa faced a number of very significant challenges prior to the arrival of Ebola. In a study published in 2015 (Loubet et al. 2015), high levels of drug resistance after first-line therapy and high levels of immunologic and virologic failure were observed. The prevalence of nucleoside reverse transcriptase inhibitors (NRTI) and non-nucleoside reverse-transcriptase inhibitors (NNRTI) resistance among the 27% of patients with detectable viral load after a median 42 months of therapy was 63% and 71%, respectively. This rapid emergence of resistance to first-line regimens jeopardizes second-line regimens, which typically have more barriers to compliance. The availability of integrase inhibitors and newer combination drugs has been extremely limited.

HIV testing initiatives, too, have met with very significant obstacles in West Africa. In a study published from Sierra Leone (Bhoobun et al. 2014) significant concerns were reported about confidentiality and stigma, and 90% of those willing to be tested expressed fears about privacy, and the majority wished to be tested at centers far from their homes because of this.

Needs in rebuilding and enhancing HIV care in Liberia and Sierra Leone after Ebola

In light of the above facts and statistics, reestablishment of facilities for the care of HIV-infected persons will require an assessment of perceived risk of EVD on the part of health care workers. The dramatic impact on staffing resulting from concerns for personal safety on all aspects of care must be addressed realistically before much progress can be made in strengthening HIV care. The approach to this vital issue must include education of workers on the effectiveness of preventive measures and on the current situation regarding Ebola incidence as well as preparations for the future detection of a resurgence of cases and more effective means of routinizing the availability of PPE.

The facts regarding HIV/AIDS care in Liberia and Sierra Leone are not substantially different from what is seen elsewhere in sub-Saharan Africa. Improving HIV testing and counseling, as well as improving strategies for entry and retention in care, is a vital goal. These processes will have to be thoroughly reviewed in order to move the West African countries in the direction of 100% treatment with effective regimens. The availability of ART drugs and the appropriate sequencing of regimens require continuous reassessment. The data noted above on the persisting issues of stigma may have a direct impact on treatment adherence, as patients sacrifice their own health in favor of privacy and concealing their diagnosis from family.

Tuberculosis

In 2015, the World Health Organization (WHO) estimated that 28% of new tuberculosis cases occurred in Africa (WHO 2015). Prior to the Ebola outbreak of 2014–2016, Liberia, Guinea, and Sierra Leone had made steady investments in tuberculosis (TB) prevention and control programs over the preceding 20 years, and the incidence of TB had fallen by 18% across Africa since the year 2000 (WHO 2015). The Ebola epidemic in West Africa, however, had a direct and negative impact on TB treatment and control efforts. Approximately the same number of individuals in Liberia, Guinea, and Sierra Leone died of TB in 2014 as those died of Ebola (Ansumana et al. 2016) in 2014 and 2015

(11,900 vs. 11,000). Almost 8000 of the TB deaths were estimated to have been impacted by Ebola (Parpia et al. 2016) and the mortality rate from TB rose in all three countries after the beginning of the Ebola epidemic (Ansumana et al. 2016).

Tuberculosis care

The impact on TB likely resulted from several factors: In the Kenema District of Sierra Leone, after two health care workers at one directly observed treatment (DOTS) TB center died of EVD, patients on therapy for TB avoided the clinic. In Liberia, DOTS centers comprised 62% of the health care facilities that were closed due to the EVD epidemic (Lori et al. 2015). In all three countries, resources for TB and other health care needs were diverted to the efforts against Ebola.

Tuberculosis prevention

The WHO advised against vaccination campaigns during the Ebola outbreak in order to reduce the threat of Ebola (UN Childrens Fund 2014). In addition, many parents avoided vaccination clinics for the fear of contamination with Ebola virus or, in some cases, conspiracy theories in which needles used to administer vaccines were contaminated with the virus. It is thought that these factors were, in part, responsible for the rise in TB cases after the Ebola epidemic (Ansumana et al. 2016).

Malaria in Children

The annual prevalence of malaria among children less than five years of age is estimated to be between 43% and 45% in Liberia, Guinea, and Sierra Leone (Parpia et al. 2016). In the sub-Saharan countries of Africa, uncomplicated malaria is typically treated without confirmatory tests (Ansah et al. 2015). For this reason estimates of the impact of the Ebola epidemic in West Africa focused on severe malaria (Parpia et al. 2016). On account of reduction in access to treatment for severe malaria resulting from the Ebola epidemic, it is estimated that an additional 4275 deaths among children less than five years of age occurred as a result of malaria in Guinea, 788 in Liberia, and 1755 in Sierra Leone, the three countries of West Africa most impacted by the Ebola epidemic (Parpia et al. 2016).

The increase in death rates from tuberculosis, severe malaria, and HIV/AIDS, which resulted from diversion of resources to combat Ebola and reluctance among the public to go to health care facilities because of fear of Ebola, together resulted in more deaths from these diseases than from Ebola in the three heavily impacted countries of West Africa. In fact, the increase in HIV, TB, and malaria deaths rivaled the total deaths from Ebola (Table 5.2).

TABLE 5.2 Increased deaths in key infectious diseases. Researchers at the Yale School of Public Health estimated that a 50% reduction in access to health care during the Ebola outbreak—a conservative estimate of the epidemic's impact on health care availability—led to about 10,000 deaths from malaria, HIV/AIDS, and tuberculosis, deaths that otherwise would not have occurred. These estimates do not count excess deaths from other causes

	INCREASED NUMBER OF MALARIA DEATHS	PERCENTAGE INCREASE IN MALARIA DEATHS	INCREASED NUMBER OF HIV/AIDS DEATHS	PERCENTAGE INCREASE IN HIV/AIDS DEATHS	INCREASED NUMBER OF TB DEATHS	PERCENTAGE INCREASE IN TB DEATHS	TOTAL INCREASED DEATHS
Guinea	4275	48.0	713	16.2	1281	51.1	6269
Liberia	788	53.6	155	13.0	592	59	1535
Sierra Leone	1755	50.0	223	9.1	841	61.4	2819
Total	6818		1091		2741		10,623

Source: Parpia, A.S. et al., Emerg. Infect. Dis., 22, 433–441, 2016.

Death rates from other conditions, both infectious and noninfectious diseases are likely to have gone up for similar reasons.

ISSUES OF MATERNAL HEALTH

By May 2015, Guinea, Liberia, and Sierra Leone, respectively, had lost 78 (1.45%), 83 (8.07%), and 79 (6.85%) of their doctors, nurses, and midwives. Researchers at the World Bank estimated that this caused an increase in maternal mortality rates in the three countries that would lead to an estimated 4022 additional deaths of women per year (Table 5.3). They added: "It is important to keep in mind that these are estimates of the direct effect of losing health personnel; these estimates do not take into account potential indirect effects such as lower health system utilization due to fear of contracting Ebola" (Evans et al. 2015). These researchers also estimated that the loss of health workers caused increase in infant and child (under-five) mortality, ranging across the three countries from 7% to 20% for infant mortality and 10% to 28% for child mortality. The infant and child estimates, however, all had 95% confidence intervals that included the effects of zero and below.

TABLE 5.3   Estimated increased maternal deaths

	MATERNAL MORTALITY RATIO (PER 100,000 LIVE BIRTHS)			NUMBER OF MATERNAL DEATHS
	PRE-EBOLA (2013)	MAY 2015	% CHANGE	ESTIMATED ADDITIONAL DEATHS PER YEAR
Guinea	650	897	38	1083
Liberia	640	1347	111	1094
Sierra Leone	1100	1916	74	1845
Total				4022

Source: Evans, D.K. et al., The next wave of deaths from Ebola? The impact of health care worker mortality. World Bank Policy Research Working Paper 7344, July 2015. http://documents.worldbank.org/curated/en/408701468189853698/pdf/WPS7344.pdf.

REFERENCES

Alves DA, Honko AN, Kortepeter MG et al. Necrotizing scleritis, conjunctivitis, and other pathologic findings in the left eye and brain of an Ebola virus-infected Rhesus macaque (*Macaca mulatta*) with apparent recovery and delayed time of death. *J Infect Dis* 2016;213P:57–60.

Ansah EK, Narh-Bana S, Affran-Bonful H et al. The impact of providing rapid diagnostic malaria tests on fever management in the private retail sector in Ghana: A cluster randomized trial. *BMJ* 2015;350:h1019. doi:10.1136/bmj.h1019.

Ansumana R, Keitell S, Roberts GMT et al. Impact of infectious disease epidemics on tuberculosis diagnostic, management, and prevention services: Experiences and lessons from the 2014–2015 Ebola virus disease outbreak in West Africa. *Int J Infect Dis* 2016;56:101–104. doi:10.1016/j.ijid.2016.10.010.

Bausch DG, Towner JS, Dowell SF et al. Assessment of the risk of Ebola virus transmission from bodily fluids and fomites. *J Infect Dis* 2007;196(Suppl 2):S142–S147.

Bhoobun S, Jetty A, Koroma MA et al. Facilitators and barriers related to voluntary counseling and testing for HIV among young adults in Bo, Sierra Leone. *J Community Health* 2014;39(3):514–520.

Bwaka MA, Bonnet MJ, Calain P et al. Ebola fever in Kikwit, Democratic Republic of the Congo: Clinical observations in 103 patients. *J Infect Dis* 1999;179(Suppl 1):S1–S7.

Castillo-Chavez C, Curtiss R, Daszak P et al. Beyond Ebola: Lessons to mitigate future pandemics. *Lancet Glob Health* 2015;3(7):e354–e355. doi:10.1016/S2214-109X(15)00068-6.

Chertow DS, Kleine C, Edwards JK et al. Ebola virus disease in West Africa—Clinical manifestations and management. *N Engl J Med* 2014;371(22):2054–2057.

Christie A, Davies-Wayne GJ, Cordier-Lasalle T et al. Possible sexual transmission of Ebola virus—Liberia, 2015. *MMWR* 2015;64:479–481.

Evans DK, Goldstein M, Popova A. The next wave of deaths from Ebola? The impact of health care worker mortality. World Bank Policy Research Working Paper 7344, July 2015. http://documents.worldbank.org/curated/en/408701468189853698/pdf/WPS7344.pdf.

IRIN. 2014. http://www.irinnews.org/report/100869/ebola-hampers-hiv-aids-care-in-liberia.

Jones KE, Patel NG, Levy MA et al. Global trends in emerging infectious diseases. *Nature* 2008;451(7181):990–993.

Kibadi K, Mupapa K, Kuvula K et al. Late ophthalmologic manifestations in survivors of the 1995 Ebola virus epidemic in Kikwit, Democratic Republic of the Congo. *J Infect Dis* 1999;179(Suppl 1):S13–S14.

Lori JR, Rominski SD, Perosky JE et al. A case series study on the effect of Ebola on facility-based deliveries in rural Liberia. *BMC Pregnancy Childb* 2015;15:254.

Loubet P, Charpentier C, Visseaux B et al. Short communication: Prevalence of HIV-1 transmitted drug resistance in Liberia. *AIDS Res Hum Retroviruses* 2014;30(9):863–866.

Loubet P, Charpentier C, Visseaux B et al. Prevalence of HIV-1 drug resistance among patients failing first-line ART in Monrovia, Liberia: A cross-sectional study. *J Antimicrob Chemother* 2015;70(6):1881–1884.

Mate SE, Kugelman JR, Nyenswah TG et al. Molecular evidence of sexual transmission of Ebola virus. *N Engl J Med* 2015;373:2448–2454.

Morse SS, Mazet JAK, Woohouse M et al. Prediction and prevention of the next pandemic zoonosis. *Lancet* 2012;380:1956–1965.

Parpia AS, Ndeffo-Mbah ML, Wenzel NS, Galvani AP. Effects of response to 2014–2015 Ebola outbreak on deaths from malaria, HIV/AIDS, and tuberculosis, West Africa. *Emerg Infect Dis* 2016;22(3):433–441.

Sagui E, Janvier F, Baize S et al. Severe Ebola virus infection with encephalopathy: Evidence for direct virus involvement. *Clin Infect Dis* 2015;61:1627–1628.

Soka MJ, Choi MJ, Baller A et al. Prevention of sexual transmission of Ebola in Liberia through a national semen testing and counselling programme for survivors: An analysis of Ebola virus RNA results and behavioural data. *Lancet Glob Health* 2016;4:e736–e743.

Tattevin P, Baysah MK, Raguin G et al. Retention in care for HIV-infected patients in the eye of the Ebola storm: Lessons from Monrovia, Liberia. *AIDS* 2015;29:N1–N2.

United Nations Children's Fund. In Sierra Leone, vaccinations another casualty of Ebola. UNICEF; 2014. http://www.unicef.org/infobycountry/sierraleone—76892.html.

Varkey JB, Shantha JG, Crozier I et al. Persistence of Ebola virus in ocular fluid during convalescence. *N Engl J Med* 2015;372(25):2423–2427.

World Bank. *The Economic Impact of the 2014 Ebola Epidemic: Short- and Medium-Term Estimates for West Africa.* Washington, DC: World Bank, 2014. doi:10.1596/978-1-4648-0438-0.

World Health Organization (WHO). Global tuberculosis report 2015. WHO.HTM/TB/2015.22. Geneva, Switzerland, 2015. http://who.int/tb/publications/global__report.

PART TWO

Science and Medicine of Ebola

Virus

6

INTRODUCTION

Considering the relative rarity of human Ebola virus infections prior to the 2014–2016 West African outbreaks, a substantial amount was already known about the virus when the outbreak occurred. Five species had been recognized and characterized. Details were known about the virus' structure, genetics, replication cycle, and means of infecting cells. This body of knowledge will be essential in the development of vaccines and targeted therapies. What follows is a brief overview of this information, with correlations drawn between steps in the viral life cycle and clinical manifestations of Ebola virus disease (EVD).

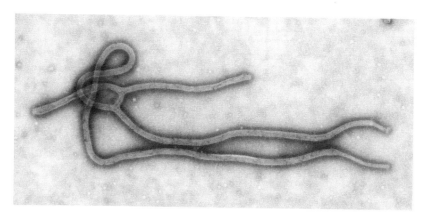

PHOTO 6.1 Ebola virus virion. An Ebola virus virion, shown in a transmission electron microscopic (TEM) image. (Courtesy of Cynthia Goldsmith, CDC, Atlanta, GA.)

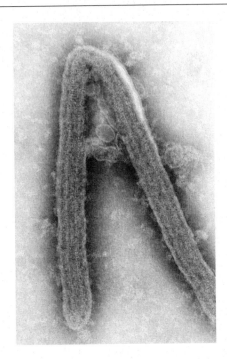

PHOTO 6.2　Marburg virus. A Marburg virus virion, shown in a transmission electron microscopic (TEM) image. Similar to Ebola, Marburg is a zoonotic RNA virus of the filovirus family. (Courtesy of F.A. Murphy, CDC, Atlanta, GA.)

VIRAL NOMENCLATURE

The family Filoviridae comprises two genera: the Ebola and the Marburg viruses. There are five recognized species in the genus Ebola. These are

> *Zaire ebolavirus* (ZEBOV): First recognized in 1976 in a teacher in Zaire (now Democratic Republic of Congo) who presented with symptoms suggestive of malaria followed by a diffuse rash as well as nausea, vomiting, and diarrhea, an illness similar to that seen in the 2014–2016 West African outbreak.
>
> *Sudan ebolavirus* (SEBOV): Also first identified in 1976 in Sudan.
>
> *Reston ebolavirus* (REBOV): Recognized in 1989 in an outbreak among macaque monkeys. REBOV has also been identified in affected animals imported from the Philippines. REBOV is closely related to SEBOV and its appearance in the Philippines is unexplained.

Cote d'Ivoire ebolavirus (CIEBOV): Isolated in 1994 from chimpanzees in the Ivory Coast.

Bundibugyo ebolavirus: Identified in Uganda in 2008.

The 2014–2016 outbreak of Ebola virus disease (EVD) was caused by the Zaire species of Ebola (ZEBOV). This species was the first recognized in association with a human outbreak of EVD in 1976 in Zaire (now the Democratic Republic of the Congo) (Johnson et al. 1977) and has been the cause of multiple outbreaks since then in Central Africa (Khan et al. 1999).

Viral Structure and Life Cycle

Ebola virus appears under electron microscopy as a pleomorphic filamentous structure, which varies in length from 300 to 1500 nm (Murphy et al. 1978) with a diameter of approximately 80 nm. The filaments, which are indistinguishable from those of Marburg virus, may form U shaped or circular structures (Bowen et al. 1977; Johnson et al. 1977; Pattyn et al. 1977; Murphy et al. 1978). The virus consists of a lipid envelope that is derived from host membranes containing glycoprotein protrusions. This envelope surrounds a matrix of VP40 and VP24 proteins and a 40–50 nm nucleocapsid containing the viral proteins VP30, VP35, NP, and L. These proteins function as follows (Sanchez et al. 1993):

Nucleoprotein (NP): Necessary for the formation of nucleocapsid structures.

VP35: Structural protein of the nucleocapsid. Necessary for replication and transcription. Also functions to block the effect of host interferon and antiviral activity.

VP24: Along with NP and VP35 forms the nucleocapsid. It binds to the plasma membrane in infected host cells and has a role in virion assembly. Also, suppresses host antiviral activity.

VP40: The most abundant viral protein. Determines viral configuration and is essential for viral budding from host cells.

VP30: A constituent of the nucleocapsid along with VP 24 and NP. Also has a role in initiation of transcription.

ZEBOV L protein: An RNA-dependent RNA polymerase and the largest protein in the virus. Together with VP35, it transcribes and replicates the genome.

GP precursor: Forms the glycoproteins (GP1, 2, soluble GP [sGP] and small soluble GP [ssGP]) and, thus, the peplomers in the viral envelope.

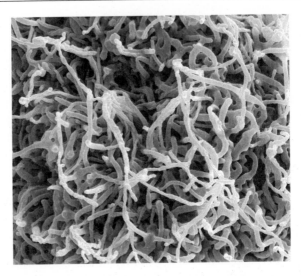

PHOTO 6.3   Ebola virus replicating. With a magnification of 50,000X, the scanning electron microscope (SEM) image developed in 2014 shows filamentous Ebola virus particles replicating from an infected VERO E6 cell. (Courtesy of the National Institute of Allergy and Infectious Diseases, Bethesda, MD.)

Viral Genome and Replication

Ebola virus is genetically similar to Marburg virus. The genetic core of Ebola virus is a negative-sense, single-stranded RNA (Regneary et al. 1980; Sanchez et al. 1993). The genome resembles those of rhabdoviruses and paramyxoviruses. The RNA is transcribed into mRNA. Viral polymerase produces an antigenome, which is used as a template for production of single-stranded RNA (Singh and Ruzek 2013).

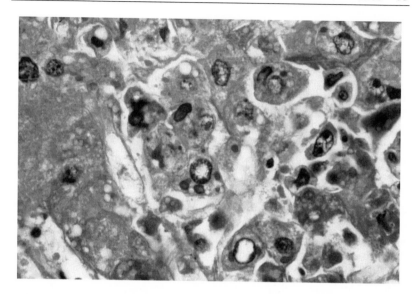

PHOTO 6.4   Infected liver cells. A micrograph from 1977 shows human liver cells infected with the Ebola virus. (Courtesy of Dr. Fred Murphy, CDC, Atlanta, GA.)

Cellular Entry

A variety of receptors have been identified on different cell types that facilitate entry by the virus. Cells initially infected are mononuclear cells. After binding to the receptor, the virus enters the cell by endocytosis. These infected cells then migrate to lymphoid tissue, including regional lymph nodes and the spleen, as well as lymphocyte aggregates in other tissues (Singh and Ruzek 2013).

Viral Budding

The virus assembles the viral proteins described above into complexes that migrate to the cell membrane where budding and release of virions takes place (Singh and Ruzek 2013).

Viral Reservoir

Despite nearly 40 years of efforts to identify the animal reservoir of Ebola virus, it has only recently been identified with a degree of certainty. In 2005, several species of fruit bats were found to contain Ebola RNA from the ZEBOV species (Leroy et al. 2005). In subsequent research (Pourrut et al. 2009), additional species of fruit bats, as well as an insectivorous species, were found to carry evidence of ZEBOV infection. The geographical areas in Africa where Ebola outbreaks have occurred all fall within the geographical areas covered by one or more of the relevant species of bat. Further evidence of the likely role of bats in Ebola transmission comes from the observation that some species can be transiently infected and have documented viremia (Swanepoel et al. 1996). However, although Marburg virus has been isolated from fruit bats and genetic evidence of the Ebola virus has been found in bats in endemic areas, intact Ebola virus has not been recovered from these animals at the time of this writing.

Routes of Transmission

Nonsexual transmission of Ebola virus from person to person is inefficient and always requires direct contact between body fluid containing substantial amounts of the virus with the skin, or most efficiently, the mucous membranes (Bausch et al. 2007). As the clinical course of EVD progresses, levels of virus in the blood and other body fluids rise to extremely high levels. Transmission through contact with the dead bodies of victims has been well documented, and represented a substantial route of transmission in the West African outbreak of 2014–2016.

Transmission from an infected pregnant woman to the fetus also occurs (see discussion in Chapter 7). Other than these three routes—direct contact, sexual transmission, and transmission from woman to fetus—no means of spreading the infection has been confirmed. However, because of the unprecedented number of cases that occurred during the 2014–2016 West African epidemic and the resultant difficulty in contact tracing, as well as fear of the spread of what was rightfully considered an extremely dangerous infection, concerns were expressed that novel, important routes of transmission might go unrecognized. Among the possibilities raised were

Airborne transmission: Although the possibility of airborne transmission has been suggested (Osterholm et al. 2015), there is no evidence that this has ever occurred. A 2016 review of studies of spread of Ebola infection within households from prior outbreaks in Africa (Dean et al. 2016) demonstrates that the risk of contagion correlates with the type and frequency of direct contact with body fluids. No direct evidence of airborne transmission was noted. Concerns that mutations could permit the virus to be transmitted by this route have been expressed, but no evidence that such mutations have occurred exists. Furthermore, a series of mutations probably would be needed for this route of transmission to become significant.

Transmission by fomite: Another concern that took a number of forms during the 2014–2016 epidemics was that Ebola infection could spread by contact with contaminated, inanimate objects such as furniture, toilets, airline seats, and the like. In a 2016 meta-analysis, Dean and colleagues (Dean et al. 2016) sought to clarify the factors associated with infection by secondary household contact. In this analysis, derived from studies conducted before the 2014–2016 epidemics, little or no transmission appeared to occur without direct physical contact. As would be expected, the extent of contact correlated with the probability of acquiring infection. In another analysis prior to the 2014–2016 epidemic, Bausch and colleagues (Bausch et al. 2007) found evidence of Ebola in 2 of 33 environmental specimens and 16 of 54 specimens of various body fluids and concluded that the risk of transmission by contact with fomites is small and can be controlled through the use of infection control practices recommended for hemorrhagic fever viruses.

SITES OF PERSISTENCE OF EBOLA VIRUS AFTER RECOVERY

Persistence in Genital Fluids/Possible Sexual Transmission

The potential for sexual transmission of Ebola virus has long been recognized, and public health officials recommended sexual abstinence after recovery from EVD. Prior to the West African outbreak, the presence of the virus in both semen and vaginal fluids, detected by polymerase chain reaction, had been demonstrated in several studies (Emond et al. 1977; Rodriguez et al. 1999; Richards et al. 2000). Prior to the recent outbreak, the longest documented period of persistence in this way was 101 days after the onset and 80 days after the clearance of viremia (Rodriguez et al. 1999), and the duration of viral shedding and potential transmissibility was not known. During and after the 2014–2016 epidemics, the period of viral persistence in genital fluids and potential transmissibility of EVD was further clarified. Far from an unusual phenomenon, viral carriage in semen was demonstrated in 100% of a small cohort of men tested within three months of the onset of clinical illness (Deen et al. 2015). Although the virus has been shown to persist in the semen of men who have recovered from EVD, sexual transmission was not documented until the recent outbreak and was thought to represent only a theoretical risk. However, in March 2015, in Liberia, a female sexual partner of a man convalescing from EVD became infected, apparently by this route (Christie et al. 2015; Mate et al. 2015). Her infection occurred 30 days after the last confirmed case in that country, suggesting that sexual transmission was the likeliest route of transmission. The man had cleared virus from his blood 155 days prior to the apparent instance of transmission (Christie et al. 2015). Although virus could not be isolated from his semen, it was detected by PCR and was genetically similar to virus isolated from the blood of the female partner (Mate et al. 2015). A dramatic example of late male-to-female transmission was described by Diallo and colleagues (Diallo et al. 2016) in which a male was found to have virus in the semen 531 days after the onset of acute illness. This resulted in sexual transmission 470 days after the onset of his symptoms and a subsequent cluster of cases in Guinea and Liberia.

Although few cases of sexual transmission have been documented at the time of this writing, the high rate of detection of virus in genital fluids, particularly semen, by PCR, in prior studies, suggests that this route may be more common than is typically assumed. Particularly in the setting of a widespread outbreak, sexual transmission may be difficult to distinguish from transmission by close, nonsexual contact. Interim guidance issued by the World Health Organization includes the recommendation that sexual contact be avoided until two semen samples are confirmed negative or six months have passed since the onset of symptoms (WHO 2015). Compliance rates with this recommendation and its practicality have not yet been determined. The maximum period of carriage of the virus has not yet been determined. However, the man in the more extreme case above had evidence of virus in his semen 531 days after acute illness began.

Persistence in Cerebrospinal Fluid and Vitreous Fluid

Two health care workers who had traveled to the endemic area during the 2014–2016 epidemics were documented to have persistent infection in two unexpected places: the central nervous system and the eye. The presence of Ebola virus was found by PCR in the cerebrospinal fluid (CSF) of a patient who had recovered from viremia (Howlett et al. 2016) in Sierra Leone. Similarly, high levels of virus were detected in the CSF more than nine months after recovery from acute EVD (BBC 2015). In the other patient, virus was isolated from the vitreous three months after recovery from acute infection. Both of these health care workers had received immunotherapy during their acute infections. This has raised the concern that this form of therapy may increase the likelihood of persistence of infection in these sites and of clinical recrudescence (Fischer and Wohl 2016).

ROLE OF FUNERAL PRACTICES IN EBOLA VIRUS TRANSMISSION

Because of the need for close contact in order for transmission to occur, outbreaks have often begun in communities and within families when contact occurs with sick individuals, including by traditional healers (Georges et al. 1999) or with dead bodies in the preparation for burial (Allaranga et al. 2010).

ASYMPTOMATIC INFECTION

Asymptomatic or minimally symptomatic infection with Ebola virus appears to occur (Leroy et al. 2000), although its frequency is not fully known. Rowe and colleagues (Rowe et al. 1999) found evidence of subclinical infection in household contacts of cases. In a review of studies published prior to the 2014–2016 West African epidemics, Dean and colleagues estimated that 27.1% of infections are asymptomatic. In the PREVAIL III study in Liberia it was found that 49% of close asymptomatic contacts of survivors were found to be IgG positive for Ebola (Fallah 2016).

Asymptomatic infection appears to carry a negligible risk of transmission by direct, nonsexual contact. However, transmission through breast milk has been suggested (Arias et al. 2016) but not confirmed and transmission by sexual contact with asymptomatic individuals recovering from acute diseases has been documented (see above), although only rarely.

The degree of protection, if any, against symptomatic disease that asymptomatic infection confers is uncertain.

REFERENCES

Allaranga Y, Kone ML, Formentay P et al. Lessons learned during active epidemiological surveillance of Ebola and Marburg viral hemorrhagic fever epidemics in Africa. *East Afr J Public Health* 2010;7:30–36.

Arias A, Watson SJ, Asogun D et al. Rapid outbreak sequencing of Ebola virus in SL identifies transmission chains linked to sporadic cases. *Virus Evol* 2016. doi:10.1093/ve/vew016.

Bausch DG, Towner JS, Dowell SF et al. Assessment of the risk of Ebola virus transmission from bodily fluids and fomites. *J Infect Dis* 2007;196(Suppl 2):S142–S147.

BBC. Ebola nurse Pauline Cafferkey "in serious condition." 2015. http://www.bbc.com/news/uk-scotland-34483584.

Bowen ET, Lloyd G, Harris WJ et al. Viral hemorrhagic in southern Sudan and northern Zaire. Preliminary studies on the aetiological agent. *Lancet* 1977;1:571–573.

Christie A, Davies-Wayne GI, Cordier-Lasalle T et al. Possible sexual transmission of Ebola virus—Liberia, 2015. *MMWR* 2015;64:479–481.

Dean NE, Halloran E, Yan Y, Longini IM. Transmissibility and pathogenicy of Ebola virus: A systematic review and meta-analysis of household secondary attack rate and asymptomatic infection. *Clin Infect Dis* 2016;62(10):1277–1286.

Deen GF, Knust B, Broutet N et al. Ebola RNA persistence in semen of Ebola virus disease survivors: Preliminary report. *N Engl J Med* 2015. doi:10.1056/NEJMoa1511410.

Diallo B, Sissoko D, Loman NJ et al. Resurgence of Ebola virus disease in Guinea linked to a survivor with virus persistence in seminal fluid for more than 500 days. *Clin Infect Dis* 2016;63(10):1353–1356.

Emond RT, Evans B, Bowen ET, Lloyd G. A case of Ebola virus infection. *Br Med J* 1977;2:541–544.

Fallah M. A cohort study of survivors of Ebola virus infection in Liberia (PREVAIL III) abstract 74LB. In: *Program and Abstracts of the 2016 Conference on Retroviruses and Opportunistic Infections*, Boston, MA, February 22–25, 2016.

Fischer WA, Wohl DA. Confronting Ebola as a sexually transmitted infection. *Clin Infect Dis* 2016;62(10):1272–1276.

Georges AJ, Leroy EM, Renaut AA et al. Ebola hemorrhagic fever outbreaks in Gabon, 1994–1997: Epidemiologic and health control issues. *J Infect Dis* 1999;179(Suppl 1):S65–S75.

Howlett P, Brown C, Helderman T et al. Ebola virus disease complicated by late-onset encephalitis and polyarthritis, Sierra Leone. *Emerg Infect Dis* 2016;22:150–152.

Johnson KM, Lange JV, Webb PA, Murphy FA. Isolation and partial characterization of a new virus causing acute hemorrhagic fever in Zaire. *Lancet* 1977;1:569–571.

Khan AS, Tshioko FK, Haymann DL et al. The reemergence of Ebola hemorrhagic fever, Democratic Republic of the Congo, 1995. *J Infect Dis* 1999;179(Suppl 1):S76.

Leroy EM, Baize S, Volchkov VE et al. Human asymptomatic Ebola infection and strong inflammatory response. *Lancet* 2000;355:2210–2215.

Leroy EM, Kumlungui B, Pourrut X et al. Fruit bats as reservoirs of Ebola virus. *Nature* 2005;438:575–576.

Mate SE, Kjugelman JR, Nyenswah TG et al. Molecular evidence of sexual transmission of Ebola virus. *N Engl J Med* 2015;373:2446–2454.

Murphy FA, Van der Groen SG, Whitfield SG, Lange JV. Ebola and Marburg virus morphology and taxonomy. In Murphy FA (ed), *Ebola Virus Haemorrhagic Fever*, 1978, pp. 61–82. Amsterdam, the Netherlands: Elsevier/North-Holland.

Osterholm MT, Moore KA, Kelley NS et al. Transmission of Ebola viruses: What we know and what we do not know. *mBio* 2015;6(2):e00137-15. doi:10.1128/mBio.00137-15.

Pattyn S, Van der Groen G, Courteille G, Jacob W, Piot P. Isolation of Marburg-like virus from a case of haemorrhagic fever in Zaire. *Lancet* 1977;1:P573–P574.

Pourrut X, Souris M, Towner JS et al. Large serological survey showing cocirculation of Ebola and Marburg viruses in Gabonese bat populations and a high seroprevalence of both viruses in *Rousettus aegyptiacus*. *BMC Infect Dis* 2009;9:159.

Regneary RL, Johnson KM, Kiley MP. Virion nucleic acid of Ebola virus. *J Virol* 1980;36:465–469.

Richards GA, Murphy S, Jobson R et al. Unexpected Ebola virus in a tertiary setting: Clinical and epidemiologic aspects. *Crit Care Med* 2000;28:240–244.

Rodriguez LL, De Roo A, Guimard Y et al. Persistence and genetic stability of Ebola virus during the outbreak in Kikwit, Democratic Republic of the Congo. *J Infect Dis* 1995;179(Suppl 1):S170–S176.

Rowe AK, Bertolli J, Khan AS et al. Clinical virologic, and immunologic follow-up of convalescent Ebola hemorrhagic fever patients and household contacts, Kikwit, Democratic Republic of the Congo. Commission de Lutte contre les Epidemies a Kikwit. *J Infect Dis* 1999;179(Suppl 1):S28–S35.

Sanchez AM, Kiley MP, Holloway BP, Auperin DD. Sequence analysis of the Ebola virus genome: Organization, genetic elements, and comparison with the genome of Marburg virus. *Virus Res* 1993;29:215–240.

Singh SK, Ruzek D. *Viral Hemorrhagic Fevers*. Boca Raton, FL: CRC Press, 2013.

Swanepoel R, Leman PA, Burt FJ et al. Experimental inoculation of plants and animals with Ebola virus. *Emerg Infect Dis* 1996;2:321–325.

World Health Organization (WHO). Interim advice on the sexual transmission of ebola virus disease. 2015. http://www.who.int/reproductivehealth/topics/rtis/ebola-virus-semen/en/.

Pathophysiology and Clinical Features of Ebola Virus Infection

7

INTRODUCTION

Prior to the West African outbreak of 2014–2016, the clinical features of Ebola virus disease (EVD) had been described in the relatively small and brief outbreaks that had occurred in remote and rural areas of Africa after the infection was first recognized in 1976. The much greater number of cases and longer duration of the West African outbreak, and the fact that it occurred in heavily populated areas, led to an enhanced understanding of the patterns of the disease. Knowledge was gained about clinical and laboratory features and response to treatment, both in Africa and in cases treated in Europe and the United States. Recent insights into pathophysiology and its relationship to the clinical manifestations of EVD will be provided in this chapter. In addition, this chapter will review postrecovery features of EVD, which had not been described in prior outbreaks, as well the newly identified possibility of long-term persistent infection and carriage.

THE PATHOPHYSIOLOGY
OF EBOLA VIRUS DISEASE

Our understanding of the pathophysiology of Ebola virus disease (EVD) is incomplete. Historically it has been a rare infection, occurring in small, sporadic outbreaks in remote areas of Central and East Africa, making research difficult into pathophysiologic processes. As a result, much insight into these processes has been derived from animal models, both primate and rodent, and has been extrapolated to human infection. The pathophysiology of EVD is complex (Martines et al. 2015) and involves dramatic changes in the immune response as well as endothelial dysfunction and coagulation disorders and direct end-organ damage seen at postmortem examination. As discussed in Chapter 6, human infection begins when infected body fluid comes into contact with the skin or mucous membranes, most likely facilitated by microscopic breaks to permit entry. Transmission of infection appears to be possible only after the onset of symptoms. The concentration of virus in the blood and body fluids rises quickly during the symptomatic phase of EVD (CDC 2015) increasing the risk of transmission through direct contact, including through contact with bodies postmortem. As reviewed in Chapter 3, "Prevention and Containment," human-to-human airborne transmission and transmission via fomites have been hypothesized but not demonstrated (CDC 2015).

Infection appears to begin with the binding of the virus by means of glycoprotein spikes within the envelope to various host cells. This is followed by internalization into the cells within endosomes (Kawaoka 2005). The virus possesses tropism for a variety of cell types and has been shown to attack cells in an asynchronous process in which it may be bound to cell membranes for an extended period before it is internalized (Reynard and Volchkov 2015). It replicates best within dendritic cells, monocytes, and macrophages (Dahlmann et al. 2015), which carry it to regional lymph nodes, the spleen, and other sites, resulting in the features of disseminated infection.

Cytokine and chemokine expression during the course of infection may differ significantly between strains of the virus. In earlier outbreaks, levels of tumor necrosis factor alpha (TNF-alpha) and interferon gamma (IFN-gamma) had been found to be increased in fatal *Zaire ebolavirus* (ZEBOV) infection (Villinger et al. 1999). But this was not the case in patients infected with the Sudan species of the virus, SEBOV (Hutchinson and Rollin 2007). High levels of several interleukins, including IL-6, IL-8, and IL-10, and the macrophage inflammatory protein-1beta were associated with fatal SEBOV infections.

The *cytokine storm* associated with EVD is similar to that seen in other overwhelming infections, including those caused by a host of bacterial, viral, fungal, and even parasitic infections. It brings with it several recognizable clinical features of the systemic inflammatory response syndrome (SIRS) including fever, tachycardia, tachypnea, and hypotension, which may be rapidly joined by such common evidence of end-organ compromise as lactic acidemia, renal insufficiency, hepatic inflammation, ischemia, hypoxemia, and central nervous system depression. This syndrome in all settings is usually accompanied by relative intravascular volume depletion resulting from peripheral vasodilation and reduced cardiac output. As discussed in the next section, in EVD, this problem is made more profound by frequently dramatic fluid losses through the gastrointestinal tract and may, rarely, be made more severe by the presence of bacterial coinfection (Kreuels et al. 2014).

CLINICAL MANIFESTATIONS OF EBOLA VIRUS DISEASE

The typical case of Ebola virus disease (EVD) begins with influenza-like features and culminates in the kind of *cytokine storm* described above. The clinical manifestations of EVD have been divided into three phases (Beeching et al. 2014). The first phase, typically lasting several days, consists of nonspecific findings, such as fever and muscle aches (Beeching et al. 2014), mimicking many other infectious diseases. The second phase typically includes gastrointestinal manifestations, particularly vomiting and diarrhea leading to dehydration. The third phase, usually beginning in the second week of illness, culminates in death or recovery and features vascular collapse and, in some cases, hemorrhagic and neurological complications. Hemorrhagic complications, with which filovirus infections such as EVD have been associated historically, were seen less frequently than expected in the 2014–2016 outbreaks. Fatal cases progress through an illness that resembles severe bacterial sepsis and septic shock, culminating in irreversible vascular collapse. In a series from Sierra Leone of 581 patients, those who died typically did so after 7–8 days of illness (Ansumana et al. 2015).

Because of the geographically restricted nature of the outbreaks that have occurred so far, the clinical suspicion of EVD relies first on a history of recent travel to a region where active transmission of EVD has been documented. This must be accompanied by a likely or possible close exposure to an individual known or suspected of having EVD. Such a history may be less relevant in endemic areas where exposure must often be assumed. However, even in endemic areas, malaria in particular can follow a similar clinical course, as can hemorrhagic viral infections such as Lassa fever, as well as typhoid, dengue, measles, leptospirosis, and a host of other infectious diseases common to Africa.

INCUBATION PERIOD

The incubation period varies from 1 to 21 days. In a report describing 106 patients with confirmed EVD in Sierra Leone, it was estimated to be between 6 and 12 days (Schieffelin et al. 2014) and was typically approximately eight days. This variability may reflect the route of transmission and the inoculum size transmitted. The potential for transmission of infection appears only after the onset of symptoms. This rather long asymptomatic period after infection has presented challenges to efforts at contact tracing and has contributed to concerns about EVD in travelers from West Africa in whom symptoms could potentially develop only after arrival at their destination outside of Africa.

THE PROGRESSION OF SYMPTOMS

Initial Manifestations

As noted above, the initial manifestations of EVD are nonspecific and may be confused with more frequently seen infectious diseases, particularly malaria. These early manifestations include fever in nearly 90% of cases (Bah et al. 2015) and frequently fatigue, diarrhea, headache, and muscle aches, as well as abdominal pain, sore throat, conjunctivitis, and vomiting (Shieffelin et al. 2014). In prior outbreaks, a generalized rash has been seen in as many as half of the cases, although this finding was far less common in the West African outbreak of 2014–2016.

Gastrointestinal Manifestations

Gastrointestinal manifestations, particularly diarrhea, usually began between days three and five in a large analysis of more than 700 patients from the West African outbreak of 2014–2016 (Chertow et al. 2014). At this point of the illness, viral shedding is typically very high, and the patient represents a high risk of transmission to others.

Hemorrhagic Complications

Bleeding complications occur in approximately one-third of patients and may include gastrointestinal bleeding, hemoptysis, epistaxis, gum bleeding, conjunctival bleeding, and bleeding at injection sites.

Skin Rash

As noted, a maculopapular rash has been reported in as many as 52% of patients in prior outbreaks but was seen much less frequently—in only approximately 5% of patients—in the 2014–2016 West African cases. A host of other infectious diseases seen in Africa may be associated with similar skin rashes. These include measles, parvovirus B19, rubella, and enterovirus among others, rendering the rash of EVD to be nonspecific and of little help in suggesting the diagnosis.

Vascular Leak

The loss of vascular integrity with resulting vascular leakage, primarily manifested as pulmonary edema and shock, is a feature of severe case (Wolf et al. 2015) and may lead to the need for ventilatory support. The lack of availability of facilities for sufficient fluid resuscitation and supported ventilation likely contributed to the substantially higher mortality rate among cases treated in Africa in comparison to those treated in the United States and Europe.

MODE OF DEATH

Death from EVD typically occurs during the second week of illness after a period of progressive volume depletion resulting from gastrointestinal fluid losses and capillary leak. Features of severe sepsis, including end-organ disorders, particularly involving the central nervous system, may be seen. Pulmonary edema may also be seen, attributable in part by the massive fluid resuscitation often needed to maintain blood pressure and renal perfusion.

Hemorrhagic complications, although not seen as frequently in the West African cases as in some previous outbreaks, may also be contributory. Hepatic necrosis, a frequent finding at postmortem examination, may manifest as well.

PHYSICAL EXAMINATION

Findings on physical examination are nonspecific during the first several days of illness. Fever, muscle tenderness, and conjunctivitis are common but, as noted, the generalized rash and clear indication of hemorrhagic complications were relatively uncommon in the 2014–2016 outbreaks in West Africa. As viremia and fluid losses progress, features of the systemic inflammatory response syndrome (SIRS) appear. These include hypertension, tachycardia, tachypnea, and fever. Progressive hypovolemia may lead to frank findings of severe dehydration, including dry mucous membranes, sunken eyes, and pronounced postural hypotension. Progressive pulmonary congestion, caused in part by efforts at fluid resuscitation, may lead to findings of generalized rales on lung examination and dullness to chest percussion if pleural effusions form.

LABORATORY FEATURES

Complete Blood Counts

White blood cells: Leukopenia with prominent depletion of lymphocytes is typical at the time of presentation. As EVD progresses, leukocytosis with a shift to the left and the presence of atypical lymphocytes becomes more common. In fatal cases, this persists until death. Thrombocytopenia is typically present in all cases and worsens throughout the illness in fatal cases. Anemia may be seen but is typically modest in the absence of significant bleeding.

Liver Function Studies

Abnormalities of transaminases are seen frequently but are typically less severe than those seen in viral hepatitis, despite the fact that hepatic necrosis is a characteristic finding in fatal cases (Murphy 1978). Jaundice is not commonly seen.

Renal Function Studies

Creatinine and blood urea nitrogen (BUN) rise as dehydration and hypotension develop, and refractory renal failure is associated with a poor outcome. Proteinuria and hematuria may also be seen. Dialysis has been utilized in some cases.

Electrolytes

Hypokalemia and reflecting gastrointestinal losses were seen in 50% of patients with Marburg virus in an outbreak in 1967 (Martini 1971).

Clotting Studies

Prolongation of prothrombin and partial thromboplastin times have been reported, and criteria for disseminated intravascular coagulation are frequently met in severe cases, with marked elevation of D-dimer (Rollin et al. 2007) predictive of a poor outcome.

Immunologic Derangements

Ebola virus preferentially targets cells of the immune system, including macrophages and dendritic cells. After viral attachment and infection of these cells, they release glycoproteins of viral origin. In this way, the virus can deregulate the dendritic cell response and interferes with their maturation and the development of an adaptive immune response. As noted above, infected macrophages overexpress proinflammatory cytokines which, if unchecked, may result in the *cytokine storm* through which infection causes capillary leak and end-organ damage (Villinger et al. 1999).

The results of routine laboratory tests in filovirus infection (Ebola and Marburg virus) were summarized by Kortepeter and colleagues in a review of published literature before the West African outbreak of 2014–2016 (Kortepeter et al. 2011). The exact frequency of specific abnormalities is not clear from the recent outbreak.

Markers of Severity, Prognosis

As noted above, a variety of proinflammatory cytokines, including interferons and tumor necrosis factors indicative of a high degree of immune activation have been associated with a fatal outcome in Zaire EVD. In addition, elevated levels of ferritin and of thrombomodulin may be associated with a worse prognosis (McElroy et al. 2014b). A variety of other markers were found to correlate with fatal outcomes in a study of patients infected with *Sudan ebolavirus* (SEBOV) in an outbreak in Uganda in 2000 (Rollin et al. 2007). D-dimer was elevated in both fatal and nonfatal cases but four-fold higher levels were seen in patients who died, most likely indicative of disseminated intravascular coagulation. Higher levels of the liver enzyme aspartate amino-transferase (AST), amylase, urea nitrogen, and creatinine and lower levels of serum albumin and calcium were also seen in fatal cases.

An analysis of 51 patients with confirmed EVD in Sierra Leone (Qin et al. 2015) found no differences in mortality between male and female patients, but a statistically greater mortality was found among patients over the age of 30 and in those presenting with extreme fatigue, diarrhea, vomiting, bleeding, or neurological symptoms.

Another study (Li et al. 2016) of 288 patients in Sierra Leone found that viral load of greater than 1 million copies per milliliter was predictive of a markedly higher mortality (odds ratio = 3.095) than those with lower levels of circulating virus. Age greater than 40 and diarrhea were also found to be poor prognostic findings in this analysis.

A higher rate of survival was also correlated with age in an analysis during the 2014–2016 epidemics by Qureshi and colleagues at the Donka National Hospital in Conakry, Guinea (Qureshi et al. 2015). Patients below the age of 35 had a significantly higher rate of survival.

TRENDS IN VIRAL STUDIES DURING THE COURSE OF INFECTION

In clinical settings, Ebola virus is detected in body fluids by means of the reverse transcription polymerase chain (RT-PCR) reaction technique. This advanced laboratory technique is not widely available in underresourced areas, a fact that makes confirmation of EVD difficult and presents challenges to early contact tracing.

Within 72 hours after the onset of symptoms, virus can be detected in the blood by polymerase chain reaction (PCR) in essentially all patients (Chertow et al. 2014; Ksiazek et al. 1999), although titers remain low in some patients for several days (Towner et al. 2004). Viral levels decline during the clinical recovery phase in survivors (Towner et al. 2004). In fatal cases, high levels in the blood persist through death, and the bodies of victims remain highly infectious. Other body fluids in which Ebola virus may be detected include saliva, stool, breast milk, tears, and seminal fluid (Bausch et al. 2007). This phenomenon is discussed below.

Early Identification and Diagnosis

Clinical diagnosis can also be difficult. As noted above, the initial clinical features of EVD closely resemble those of several more common infections, particularly malaria. For this reason, clinical suspicion is paramount in the initial identification of EVD, and this may be low outside of an outbreak setting.

Epidemiologic Features

Presumed close contact with individuals known or suspected to have symptomatic EVD is the first step toward establishing the clinical diagnosis both in endemic and nonendemic settings. In the regions of West Africa affected in the 2014–2016 outbreaks, such direct contact could be confirmed in household contacts or inferred through potential exposure in health care or burial settings. Many countries outside of the endemic area established screening protocols at airports and other ports of entry and in health care settings to rapidly identify individuals with recent travel to the affected countries in

West Africa and to assess them for potential contact with an infected individual and for symptoms possibly caused by EVD. The use of extensive personal protective equipment by health care personnel and specific isolation strategies are recommended for the clinical assessment if patients are suspected of having symptoms of EVD. This is discussed in detail in Chapter 3 and the Appendix.

Key Issues in Management

The clinical manifestations that, typically, must be addressed in any therapeutic plan are briefly summarized below. Chapter 8 provides more detailed discussion of management of patients with EVD.

Supportive care

The care of patients with EVD is primarily supportive. Although antiviral therapy with several agents, as well as immunotherapy using serum from convalescing patients, have both been incorporated into the treatment of several patients, the effectiveness of these strategies has not been established. The most important treatment strategy appears to be adequate fluid and electrolyte replacement. In fact, the availability of adequate supplies of intravenous fluid and electrolyte solutions is a likely explanation for the better survival of patients treated in developed countries in the recent West African outbreak.

Fluid and electrolyte losses

Loss of body fluids through the gastrointestinal tract and diaphoresis and the resultant vascular collapse poses the gravest danger in EVD. Electrolyte losses, particularly potassium, magnesium, calcium, and bicarbonate, may be rapid and profound and further complicate management of the disease. As in other hypovolemic states, lactic acidosis supervenes, and systemic acidemia caused by this, as well as by severe diarrhea, may result in potentially lethal cardiac arrhythmias. It is likely that the greater success in management of EVD in centers in the United States and in Europe primarily reflects nothing more than the enhanced capacity to address these issues. Specific strategies for fluid and electrolyte replacement are discussed in Chapter 8.

Respiratory compromise

Respiratory involvement requiring supplemental oxygen and, possibly, mechanical ventilation, is seen frequently in severe disease and may be exacerbated by progressive acidemia resulting from poor organ perfusion and gastrointestinal fluid losses. The adult respiratory distress syndrome (ARDS) and respiratory bacterial superinfection may complicate the clinical management.

Renal failure

Acute renal failure may result from hypovolemic shock as well as other systemic insults. Renal replacement therapy is discussed in Chapter 8.

Convalescence, late manifestations, and persistence of infection

Following recovery from EVD, chronic nonspecific sequelae, including fatigue, weight loss, headache, migratory arthralgias, hair loss, skin change, and anemia have been described (Kortepeter et al. 2011). In data from an outbreak in Kikwit, Democratic Republic of Congo, persistence of virus was detected in semen as long as 82 and 91 days after the onset of symptoms in one report (Rodriguez et al. 1999; Rowe et al. 1999) but not in other body fluids, including sweat, feces, urine, saliva, or vaginal secretions in specimens collected between 12 and 157 days after illness (Rowe et al. 1999). However, analysis of persistence of detectable virus in patients during the 2014–2016 West Africa outbreaks revealed viral RNA in semen after 199 days in one survivor (Christie et al. 2015) in a case of possible sexual transmission. In a study of 429 survivors in Liberia, 24 (6%) had semen that tested positive more than 12 months after the acute infection. The longest time interval between acute infection and detection of persistent Ebola virus in the semen was 565 days (Soka et al. 2016).

Symptomatic orchitis may develop several weeks after acute infection. A variety of studies of several outbreaks (Rodriguez et al. 1999; Bausch et al. 2007; Moreau et al. 2015) have documented virus between 6 and 82 days after disease onset in various body fluids including vaginal swabs, rectal swabs, urine, saliva, and breast milk.

EYE INVOLVEMENT

Ebola virus has been detected in the aqueous humor during convalescence, nine weeks after clearance of viremia. In addition, of more than 2700 survivors in Sierra Leone who were screened for ocular complications, approximately one in five were found to have uveitis (Cancedda et al. 2016). In the case of one American physician who had worked in an Ebola endemic area in 2014, uveitis was diagnosed with evidence of persistent Ebola virus in the aqueous fluid (Varkey et al. 2015).

CENTRAL NERVOUS SYSTEM INVOLVEMENT

Concerns have been expressed regarding the possibility of persistent involvement of the central nervous system. A Scottish nurse who had contracted EVD in Sierra Leone in early 2015 was diagnosed with meningitis in October 2015, which was thought to represent a recurrence. She was treated successfully, but was hospitalized again with concern regarding a possible second recurrence in early 2016 (Gulland 2016). Although Ebola virus has been detected in the cerebrospinal fluid (CSF) of only a small number of patients, the possibility of a specific Ebola encephalitis virus has been suggested (DeBreslan et al. 2016) on the basis of neurological abnormalities detected in patients with positive PCR for Ebola virus of the CSF. This may be characterized by gait instability, dizziness, aggressiveness, intolerance to frustration, and slowness of ideation (de Greslan et al. 2016).

SPECIAL POPULATIONS

Pregnant Women

Since the majority of female victims of EVD are of childbearing age, infection in pregnancy has been seen occasionally in past outbreaks and was seen in a small number of cases in the West African outbreak of 2014–2016.

Maternal–fetal transmission

Transplacental fetal infection appears to be common, and Ebola virus has been detected in placental tissue and in amniotic fluid (Baggi et al. 2014).

Incidence and outcomes in pregnancy

Through 2014, 103 cases of EVD in pregnancy had been identified in the medical literature by Bebell and Riley (2015). In this analysis, maternal death occurred in over 90%, suggesting that EVD is more severe in pregnancy. Pregnancy ended in spontaneous abortion in 32 cases. All live births resulted in infant deaths by day 19.

Clinical manifestations and management in pregnancy

The signs and symptoms of EVD in pregnant women are similar to those seen in other patients. As in other patients, diarrhea and vomiting and the attendant fluid and electrolyte losses are expected and typically precede vascular collapse in fatal cases. Hemorrhagic manifestations are occasionally seen. Fluid resuscitation may require as much as 10 liters of fluid per day. Antidiarrheals and antiemetics can be given safely (Bebell and Riley 2015). The use of fetal monitoring is of unknown value. Caesarian delivery may be hazardous in the advanced stages of EVD and, since transplacental transmission is likely to have already occurred, may not prevent fetal infection and nearly certain demise. The possibility of infection, or coinfection, with other agents that pose significant hazards in pregnancy (Brabin 1985; Foster 1996), including malaria, *Salmonella typhi*, Lassa fever, hepatitis, and others, should be considered, and empiric treatment for these infections or broad-spectrum antimicrobial therapy for bacterial sepsis should be considered in pregnant women, as in all patients with EVD. Since Ebola virus has been found in breast milk after recovery from infection (CDC 2016), breastfeeding should be discouraged until more is known about this potential route of transmission.

As in all clinical situations involving EVD, health care workers caring for pregnant women are at risk of acquiring infection themselves. This may be particularly likely during delivery and if women present with significant genital bleeding (Mupapa et al. 1999). Appropriate personal protective equipment should be used during all patient contact (see Chapter 3).

Asymptomatic infection in pregnancy

Ebola virus was detected on routine blood testing by reverse-transcription polymerase chain reaction in an asymptomatic pregnant woman presenting during her third trimester in a 2015 report from Liberia (Akerlund et al. 2015). She denied any exposure to individuals with EVD. Three days later, she became febrile. She died on day seven with symptoms of EVD. Although the full significance of this case is not clear, concern that pregnancy can conceal or mimic symptoms of early EVD is raised.

Children

Incidence

As with many aspects of EVD, clinical observations and data in children are incomplete because outbreaks have occurred in poorly resourced areas. It is thought likely that the West African outbreak of 2014–2016 could be traced to a two-year-old boy in Guinea who acquired the infection in December 2013 (Baize et al. 2014). Nonetheless, the incidence of infection in young children has appeared to be relatively low both in the West African and prior outbreaks. In the 2014–2016 cases, approximately 14% of cases occurred in children younger than 15 (WHO Ebola Response Team 2014). Even lower rates of childhood infection were seen in the 1995 outbreak in Zaire, where the incidence in those under 18 was 9% (Dowell 1996). This low attack rate in children is surprising since median age of the population is typically much lower in developing countries than in the developed world. In fact, at the time of the 1995 outbreak, more than 50% of the population was below the age of 16 (Dowell 1996). In addition, a variety of infections, including measles and varicella, are seen much more frequently in children than in adults, who have gained immunity following childhood infection. Although the reasons for a lower attack rate by EVD in children are not fully understood, it is likely that cultural practices in the 1995 outbreak, in which children were intentionally protected from exposure to sick and dying adults played a role (Dowell 1996). This may account for the similarly low incidence among children in the West African outbreak.

The relative scarcity of pediatric cases in most outbreaks has resulted in relatively little understanding of the host response in children. However, in the 2000–2001 outbreak of Sudan-associated EVD in Uganda, there were 55 laboratory-confirmed cases, which permitted an evaluation of various biomarker levels and the possibility of correlating specific immunologic responses with survival (McElroy et al. 2014a). In this study, findings indicative of a vigorous endothelial response were associated with death in children but not in adults. In contrast, children with high levels of the chemokine regulated on activation, normal T-cell expressed and secreted marker (RANTES) and lower levels of plasminogen (McElroy et al. 2014a) had greater survival. These markers were not predictive of the outcome in adults in this study.

Clinical manifestations

The initial manifestations of EVD in children appear to be similar to those in adults, that is, fever, headache, myalgia, and abdominal pain with progression to diarrhea, vomiting, and in the minority of cases, hemorrhagic manifestations (Peacock et al. 2014). In adults, these symptoms are common to a host of other infectious diseases (e.g., malaria, typhoid, and measles) seen in developing countries, a fact that increases the difficulty of diagnosis without a clear exposure history. During the first nine months of the West African outbreak of 2014–2016, the case fatality rate among children under the age of 15 was 73.4% compared to 66.1% for those between 15 and 44 and 80.4% for those 45 and older (WHO Ebola Response Team 2014).

Special considerations in neonates

Neonates born to women with EVD seldom survive longer than several weeks (Mupapa et al. 1999; Francesconi et al. 2003). To date, no healthy infants are known to have been born to mothers with symptomatic EVD (CDC 2016), although little is known about the potential for survival and appropriate treatment when modern medical care is available (CDC 2016). As noted above, in utero transmission of Ebola virus to the developing fetus appears to occur (Baggi et al. 2014). In addition, since the virus has been detected in breast milk (CDC 2016), it is possible that transmission occurs before, during, or after delivery.

Infants born to women with confirmed EVD

Infants born to women with confirmed EVD should be regarded as infected themselves and should be separated from their mothers and placed in an appropriate isolation unit. If the baby is stable, routine neonatal care including immunizations and noninvasive screening tests for congenital heart disease, hearing, and so on, should take place. The persistence of Ebola virus in breast milk has been documented (CDC 2016). Since it is unclear how long this lasts and how likely it is to transmit the virus to the infant, these infants should not be breastfed even if the mother recovers. Breast milk in such a case should be disposed of as contaminated waste. Health care workers involved in the treatment of such infants should wear full recommended personal protective equipment (PPE) (CDC 2016) for 21 days following delivery. It is recommended that circumcision and invasive blood test, including for EVD, be deferred until 21 days of life (CDC 2015). The most effective treatment measures and the most definitive way to exclude EVD in neonates have not been fully developed, although maintaining fluid and electrolyte balance would, as in adults, presumably be a key component. After 21 days, the child can be considered for discharge if blood tests are negative for Ebola infection (CDC 2016).

Infants born to women with possible EVD
If the mother is a person under investigation (PUI) for EVD, the neonate should be separated from her and isolated with appropriate precautions (CDC 2016), and breastfeeding should not occur. If EVD in the mother is ruled out, routine care of the neonate and breastfeeding can proceed and the infant can be returned to the mother.

Infants born to asymptomatic women with possible exposure to EVD
Neonates born to mothers who have potentially been exposed to Ebola virus infection should be considered to be in the same risk category as their mother. They can remain with their mother unless she develops symptoms consistent with EVD. In this case they should be separated from the mother and placed under appropriate isolation and precautions should be taken for 21 days following the last mother–child contact (CDC 2016).

DIFFERENTIAL DIAGNOSIS

A number of infectious diseases that are endemic to Africa can closely resemble EVD. Confirmation of these disorders usually requires blood tests or cultures and may be delayed in patients in whom EVD is being considered because of the possible transmission of Ebola infection to phlebotomists or laboratory personnel. Under some circumstances, such as when malaria or typhoid are suspected, empiric therapy for these infections may be considered if the diagnosis is likely to be significantly delayed. Among the most important of these diseases are

> *Malaria*: The symptoms of malaria, an infection that is quite common in many regions of Africa—in fact far more common than EVD—are comparable to those of early EVD. The abrupt onset of high fever, rigors, and myalgias often accompanied by neutropenia, thrombocytopenia, and elevated transaminases are common to both. Malaria caused by *Plasmodium falciparum* may progress to cause severe central nervous system and renal involvement as can be seen in EVD. Although the gastrointestinal symptoms and hemorrhagic

complications common to EVD after the first few days of illness would not be expected in malaria, the failure to diagnose malaria within the early stages of illness can have devastating consequences. The diagnosis of malaria, particularly in cases severe enough to mimic EVD, may almost always be made on examination of blood smears, which reveal the characteristic parasitic forms.

Lassa fever: Lassa fever is a viral hemorrhagic illness caused by the Lassa fever virus. Diagnosis may represent a significant challenge even to senior physicians practicing in endemic areas (Dahmane et al. 2014; Eze et al. 2014; Olowookers et al. 2014). It has many features in common with EVD and occurs in several countries of West Africa, including Liberia, Sierra Leone, Guinea, Nigeria, and occasionally, neighboring areas (Cross et al. 2014). It is much more common than EVD, with annual estimates of cases in these countries between 300,000 and 500,000. Most infections are mild and do not require hospitalization, but severe hemorrhagic fever occurs in approximately 5% of cases and is associated with case fatality rates of 15%–25% (CDC 2014) among hospitalized patients, although the overall mortality is only approximately 1%. Nonspecific symptoms including fever, sore throat, chest, back, and abdominal pain are seen frequently. Severe gastrointestinal fluid losses are uncommon, although vomiting may be seen, particularly in fatal cases. Hemorrhagic complications are seen in less than 20% of cases. As in EVD, the severity of symptoms typically reaches its maximum during the second week of illness. Although management of Lassa fever is typically supportive, specific antiviral therapy with ribavirin may be effective in treatment as well as for postexposure prophylaxis (Bausch et al. 2010). Blood tests are needed to distinguish between Lassa fever and EVD.

Typhoid: Typhoid fever caused by *Salmonella typhi*, similar to malaria, is common in Africa and begins as an undifferentiated febrile illness similar to EVD with symptoms that include fever and abdominal pain. Relative leukopenia and elevated transaminases are also seen frequently. Although the profound diarrhea and accompanying intravascular volume depletion common in EVD are seen infrequently in patients with typhoid, a picture of severe sepsis with hypotension and end-organ involvement may complicate distinguishing between typhoid and EVD. Clotting disorders are rare in typhoid, and the associated rash (i.e., Rose spots) does not resemble the diffused eruption sometimes seen in EVD. Confirmation of the diagnosis typically requires cultures of blood, stool, or urine for *Salmonella typhi*.

Measles (*Rubeola*): Before the onset of characteristic mucocutaneous symptoms, measles is also associated with nonspecific symptoms, which may include fever, myalgias, and arthralgias. During this phase, the diagnosis may be overlooked if EVD is a significant concern. The appearance of more typical symptoms, including cough, coryza, and most importantly, the generalized rash that begins on the face, as well as Koplik spots on the buccal mucosa, may help make the distinction between measles and EVD. However, conjunctivitis, which is seen in a significant proportion of individuals with EVD and the morbilliform rash of measles, may continue to make differentiation difficult. Severe gastrointestinal symptoms, hypovolemic shock, and hemorrhagic complications are rare in measles. Although there is no specific treatment for measles, the failure to make an early distinction between measles and EVD could result in ineffective precautions and foster the transmission of measles by the airborne route.

Meningococcal disease: Infection with *Neisseria meningitides*, which is seen commonly in some areas of Central Africa, may pose a significant diagnostic challenge when EVD is under consideration. The early symptoms, for example, fever, myalgias, and arthralgias, may mimic those caused by Ebola virus infection. Later manifestations may include bleeding diathesis and a generalized rash, both similar to those seen in EVD. The failure to recognize meningococcal infection promptly can lead to grave consequences as clinical progression may occur over hours. Early specific antibiotic therapy is essential. The public health implications of meningococcal infection are extremely significant. Potential sexual, household or health care facility contacts must be identified for potential antibiotic prophylaxis as well as to develop vaccination strategies. If meningococcal infection is suspected on clinical or epidemiological grounds, specific therapy should be instituted promptly.

Leptospirosis: Leptospirosis, an infection caused by several species of the spirochete Leptospira, is endemic throughout much of the tropical and subtropical regions of the world. The organism is often present in rodent urine. Human infection may occur after exposure via cuts and abrasions to water contaminated with Leptospira organism. The illness begins with nonspecific symptoms of fever, joint and muscle pain, headache, nausea and vomiting, and conjunctival suffusion. In severe cases there may be progression to fulminant hepatitis, renal failure, and pulmonary hemorrhage with features of severe sepsis. Death is caused by multiorgan failure. Thus severe leptospirosis may be indistinguishable from EVD on clinical grounds and early diagnosis is challenging. In some cases, after initial remission, recrudescent illness occurs with central nervous system involvement. Diagnosis is made by culture or serological means. Treatment is with penicillin derivatives or doxycycline. *Cholera*: Cholera is a diarrheal illness caused by the bacterium *Vibrio cholerae*. The illness occurs sporadically as well as in large-scale outbreaks periodically in many areas of the developing world. It could be confused with EVD during the period of massive diarrhea and resultant fluid and electrolyte imbalance and vascular collapse. Diagnosis, particularly outside of an outbreak setting, requires isolation of the organism from stool, which may be beyond the capacity of laboratories in resource-deprived areas. Treatment is with fluid and electrolyte replacement and antibiotics.

REFERENCES

Akerlund E, Prescott J, Tampellini L. Shedding of Ebola virus in an asymptomatic pregnant woman. *N Engl J Med* 2015;372(25):2467–2468.

Ansumana R, Jacobsen KH, Sahr F et al. Ebola in Freetown Area, Sierra Leone— A case study of 581 patients. *N Engl J Med* 2015;372(6):587–588.

Baggi FM, Taybi A, Kurth A, Van Herp M, Di Caro A, Wölfel R, Günther S, Decroo T, Declerck H, Jonckheere S. Management of pregnant women infected with Ebola virus in a treatment centre in Guinea, June 2014. *Euro Surveill* 2014;19(49):pii=20983. Available online: http://www.eurosurveillance.org/ ViewArticle. aspx?ArticleId=20983.

Bah EI, Lamah M, Fletcher T et al. Clinical presentation of patients with Ebola virus disease in Conakry, Guinea. *N Engl J Med* 2015;372(1):40–47.

Baize S, Pannetier D, Oestereich L et al. Emergence of Zaire Ebola virus disease in Guinea. *N Engl J Med* 2014;371(15):1418–1425.

Bausch DG, Hadi CM, Khan SH, Lertora JJL. Review of the literature and proposed guidelines for the use of oral ribavirin as postexposure prophylaxis for Lassa fever. *Clin Infect Dis* 2010;51(12):1435–1441.

Bausch DG, Towner JS, Dowell SF et al. Assessment of the risk of Ebola virus transmission from bodily fluids and fomites. *J Infect Dis* 2007;196(Suppl 2):S142–S147.

Bebell LM, Riley LE. Ebola virus disease and Marburg disease in pregnancy. *Obstet Gynecol* 2015;125:1293–1298.

Beeching NJ, Fenech M, Houlihan CF. Ebola virus disease. *BMJ* 2014;349:1–15.

Brabin BJ. Epidemiology infection in pregnancy. *Rev Infect Dis* 1985;7:579–603.

Cancedda C, Davis SM, Dierberg KL et al. Strengthening health systems while responding to a health crisis: Lessons learned by a nongovernmental organization during the Ebola virus disease epidemic in Sierra Leone. *J Infect Dis* 2016;214(Suppl 3): S153–S163. doi:10.1093/infdis/jiw345.

Centers for Disease Control and Prevention (CDC). Lassa fever: Signs and symptoms. 2014. http://www.cdc.gov/vhf/lassa/symptoms/index.html.

Centers for Disease Control and Prevention (CDC). Review of human-to-human transmission of Ebola virus. 2015. http://www.cdc.gov/vhf/ebola/transmission/human-transmission.html-sixteen.

Centers for Disease Control and Prevention (CDC). Care of a neonate born to a mother who is confirmed to have Ebola, is a person under investigation, or has been exposed to Ebola. 2016. http://www.cdc.gov/vhf/ebola/healthcare-us/hospitals/neonatal-care.html.

Chertow DS, Kleine C, Edwards JK. Ebola virus disease in West Africa—Clinical manifestations and management. *N Engl J Med* 2014;371(22):2054–2057.

Christie A, Davies-Wayne GJ, Cordier-Lasalle T et al. Possible sexual transmission of Ebola virus—Liberia, 2015. *MMWR* 2015;64(17):479–481.

Cross RW, Geisbert JB, Folarin OA et al. Lassa fever in post-conflict Sierra Leone. *PLoS Negl Trop Dis* 2014;8(3):e2748.

Dahlmann F, Biedenkopf N, Babler A et al. Analysis of Ebola virus entry into macrophages. *J Infect Dis* 2015;212(Suppl 2):S247–S257.

Dahmane A, van Griensven J, Van Herp M et al. Constraints in the diagnosis and treatment of Lassa fever and the effect on mortality in hospitalized children and women with obstetric conditions in a rural district hospital in Sierra Leone. *Trans R Soc Trop Med Hyg* 2014;108(3):126–132.

DeBreslan T, Billot M, Rousseau C et al. Ebola virus-related encephalitis. *Clin Infect Dis* 2016;63:1076–1078.

Dowell SF. Ebola hemorrhagic fever: Why were children spared? *Peds Infect Dis J* 1996;15(3):189–191.

Eze KC, Salami TA, Kpolugbo JU. Acute abdominal pain in patients with Lassa fever: Radiological assessment and diagnostic challenges. *Niger Med J* 2014;55(3):195–200.

Foster SO. Malaria in the pregnant African woman: Epidemiology, practice, research, and policy. *Am J Trop Med Hyg* 1996;55:1–3.

Francesconi P, Yout Z, Declich S et al. Ebola hemorrhagic fever transmission and risk of contacts, Uganda. *Emerg Infect Dis* 2003;9(11):1430.

Gulland A. US nurse who contracted is readmitted to hospital. *BMJ* 2016;352:1134.

Hutchinson KL, Rollin PE. Cytokine and chemokine expression in humans infected with Sudan Ebola Virus. *J Infect Dis* 2007;196:S357–S363.

Kawaoka Y. How Ebola infects cells. *N Engl J Med* 2005;352(25):2645–2646.

Kortepeter MG, Bausch DG, Bray M. Basic clinical and laboratory features of filoviral hemorrhagic fever. *J Infect Dis* 2011;204(Suppl 3):S810–S816.

Kreuels B, Wichmann D, Emmerich P et al. A case of severe Ebola virus infection complicated by gram-negative septicemia. *N Engl J Med* 2014;371(25):2394–2401.

Ksiazek TG, Rollin PE, Williams AJ et al. Clinical virology of Ebola hemorrhagic fever (EHF): Virus, virus antigen, and IgG and IgM antibody findings among EHF patients in Kikwit Democratic Republic of the Congo, 1995. *J Infect Dis* 1999;179(Suppl 1):S177–S187.

Li J, Duan HJ, Chen HY et al. Age and Ebola viral load correlate with mortality and survival time in 288 Ebola virus disease patients. *Int J Infect Dis* 2016;42:34–39.

Martines, RB, Ng DL, Greer P, Rollin PE, Zaki SR. Tissue and cellular tropism, pathology and pathogenesis of Ebola and Marburg viruses. *J Pathol* 2015;235:153–174.

Martini GA. *Marburg Virus Disease.* New York: Springer Verlag, 1971; pp. 1–9.

McElroy AK, Erickson BR, Flietstra TD et al. Biomarker correlates of survival in pediatric patients with Ebola virus disease. *Emerg Infect Dis* 2014a;20(10):1683–1690.

McElroy AK, Erickson BR, Flietstra TD et al. Ebola hemorrhagic fever: Novel biomarker correlates of clinical outcome. *J Infect Dis* 2014b;210:558–566.

Moreau M, Spencer C, Gozalbes JG et al. Lactating mothers infected with Ebola virus: EBOV RT-PCR of blood only may be insufficient. *Euro Surveill* 2015;20(3): pii=21017. http://www.eurosurveillance.org/ViewArticle.aspx?ArticleId=21017.

Mupapa K, Mukundu W, Bwada MA et al. Ebola hemorrhagic fever and pregnancy. *J Infect Dis* 1999;179(Suppl 1):S11–S12.

Murphy FA. Pathology of Ebola virus infection. In: Pattyn S, ed. *Ebola Virus Haemorrhagic Fever.* Amsterdam, the Netherlands: Elsevier, 1978; pp. 39–53.

Olowookers SA, Fatiregun AA, Gbolahan OO, Adepoju EG. Diagnositic proficiency and reporting of Lassa fever by physicians in Osun State of Nigeria. *BMC Infect Dis* 2014;14:344–352.

Peacock G, Uyeki TM, Rasmussen SA. Ebola virus disease and children: What pediatric health care professionals need to know. *JAMA Pediatr* 2014;168(12):1087–1088.

Qin E, Jingeng B, Zhao M et al. Clinical features of patients with Ebola virus disease in Sierra Leone. *Clin Infect Dis* 2015;61:491–502.

Qureshi AI, Chughtai M, Bah EI et al. High survival rates and associated factors among Ebola virus disease patients hospitalized at Donka National Hospital, Conakry, Guinea. *J Vasc Interv Neurol* 2015;8(1.5):S4–S11.

Reynard O, Volchkov VE. Entry of Ebola virus is an asynchronous process. *J Infect Dis* 2015;212(Suppl 2):S199.

Rodriguez LL, De Roo A, Guimard Y et al. Persistence and genetic stability of Ebola virus during the outbreak in Kikwit, Democratic Republic of the Congo, 1995. *J Infect Dis* 1999;179(Suppl 1):S170–S176.

Rollin PE, Bausch DG, Sanchez A. Blood chemistry measurements and D-dimer levels associated with fatal and nonfatal outcomes in humans infected with Sudan Ebola virus. *J Infect Dis* 2007;196:S364–S371.

Rowe AK, Bertolli J, Khan AS et al. Clinical, virologic and immunologic follow-up of convalescent Ebola hemorrhagic fever patients and their household contacts, Kikwit, Democratic Republic of the Congo. *J Infect Dis* 1999;179(Suppl 1): S28–S35.

Schieffelin JS, Shaffer JG, Goba A et al. Clinical illness and outcomes in patients with Ebola in Sierra Leone. *N Engl J Med* 2014;371(22):2092–2100.

Soka MJ, Choi MJ, Baller A et al. Prevention of sexual transmission of Ebola in Liberia through a national semen testing and counselling programme for survivors: An analysis of Ebola virus RNA results and behavioural data. *Lancet Glob Health* 2016;4:e736–e743.

Towner JS, Rollin PE, Bausch DG et al. Rapid diagnosis of Ebola hemorrhagic fever by reverse transcription-PCR in an outbreak setting and assessment of patient viral load as a predictor of outcome. *J Virol* 2004;78(8):4330–4341.

Varkey JB, Shantha JG, Crozier I et al. Persistence of Ebola virus in ocular fluid during convalescence. *N Engl J Med* 2015;372(25):2423–2427.

Villinger F, Rollin PI, Brar SS et al. Markedly elevated levels of interferon (IFN)-gamma, IFN-alpha, Interleukin (IL)-2, IL1- and tumor necrosis factor-alpha associated with fatal Ebola virus infection. *J Infect Dis* 1999;179(Suppl 1):S188–S191.

WHO Ebola Response Team. Ebola virus disease in West Africa: The first 9 months of the epidemic and forward projections. *N Engl J Med* 2014;371:1481–1495.

Wolf T, Kann G, Becker S et al. Severe Ebola virus disease with vascular leakage and multiorgan failure: Treatment of a patient in intensive care. *Lancet* 2015;385:1428–1435.

Supportive, Antiviral, and Immune Therapy of Patients with Ebola Virus Disease

8

INTRODUCTION

At the time of this writing, no medications have been approved for the treatment of Ebola virus diseases (EVD). One vaccine has shown promise in human trials, as described in Chapter 9. It has been emphasized that medical management should focus on adequate fluid and electrolyte replacement and that, when this is provided, survival improves substantially (Bah et al. 2015). This supportive therapy has been pointed to as, perhaps, the major reason why a higher proportion of patients treated in the United States have survived compared to those treated in West Africa.

Although a small number of patients have been treated with antiviral medications, it is not yet clear that antiviral therapy was helpful in any of the surviving patients treated so far (see below). Nonetheless, the development of potential antiviral agents has accelerated since the outbreak of 2014–2016. The development of new agents has focused both on antiviral chemotherapy and on the use of monoclonal antibodies directed at various constituents of the virus.

Ebola virus appears to produce EVD only in humans and in nonhuman primates. This fact presents challenges in the clinical evaluation of antiviral therapy as well as vaccines. Further complicating research into therapeutics is the fact that working with the virus to demonstrate effectiveness of specific therapeutic agents requires the facilities of a biosafety level 4 containment laboratory and properly trained workers.

When possible, patients known or suspected of having EVD should be treated in specialized centers with appropriate containment facilities to prevent transmission to patients and to health care workers.

SUPPORTIVE CARE

The mainstay of supportive care of patients with EVD is replacement of gastrointestinal fluid and electrolyte losses. Early in the course of illness or in milder cases, this may be accomplished by oral rehydration. However, in cases in which fluid losses are severe, intravenous therapy is necessary. The lack of availability of fluid and electrolyte solutions has been pointed to as a likely explanation for the higher mortality rates seen among patients treated in West Africa as opposed to those transported to Europe or the United States for management.

INTRAVASCULAR VOLUME REPLETION

In the most severe period of illness, typically toward the end of the first week after the onset of symptoms and through the second week, fluid losses from diarrhea and vomiting may reach 5–10 liters per day. The accompanying losses of key electrolytes, including potassium, calcium, magnesium, and sodium may contribute further to neurological dysfunction and cardiovascular complications and death. In settings where invasive monitoring is feasible, careful measurement of fluid output as well as central venous pressure and oxygen saturation are appropriate.

ANTIDIARRHEAL AGENTS

Gastrointestinal involvement in EVD is present in most cases and is severe in fatal cases. The typically massive fluid losses caused by diarrhea frequently result in profound hypovolemia and also pose a risk to caregivers since fecal material contains large amount of virus. This must be addressed with vigorous rehydration and electrolyte replacement. The potential role of antidiarrheal agents has not been systematically studied. However the antimotility, antidiarrheal agent loperamide may offer a particular benefit, in part because of its additional antiinflammatory effect (Chertow et al. 2015). This approach to therapy may be particularly beneficial in resource-poor settings where maintaining adequate fluid and electrolyte replacement in the setting of unrelenting diarrhea may be impossible.

RENAL REPLACEMENT THERAPY

Acute kidney injury results from intravascular volume depletion and is an important contributor to mortality (Wolf et al. 2015). The lack of availability of renal replacement therapy (RRT), including hemodialysis, greatly complicates the treatment of patients in resource-deprived areas. One important concern is whether hemodialysis can be provided in a manner that does not place health care workers at undue risk. Two nurses involved in the care of a Liberian man with EVD who reportedly underwent hemodialysis in a Dallas hospital contracted Ebola virus infection. Although it is not clear how infection was transmitted in this setting, hemodialysis, a procedure associated with the transmission of bloodborne pathogens such as hepatitis B and C, must be considered as a potential risk (Wolf et al. 2015). The optimal means of providing RTT is not clear. Insertion of intravascular or peritoneal dialysis catheters poses risk of transmission to the operator, perhaps made greater by the need for use of extensive personal protective equipment, which may interfere with dexterity. Insertion of peritoneal catheters may carry high risk of bowel perforation since patients may develop abdominal and bowel distension. Concerns regarding the safe use of anticoagulation and about the adequacy of current dialysis machine disinfection procedures have also been raised (Wolf et al. 2015).

ADJUNCTIVE ANTIMICROBIAL THERAPY

Since the features of severe EVD are largely indistinguishable from the severe sepsis syndrome associated with bacterial infection, malaria, and other disorders, empiric antimicrobial therapy may be indicated when definitive diagnosis of other infections is not feasible, either because of the unavailability of diagnostic tests or because of concern about contagion resulting from phlebotomy or specimen handling. Broad spectrum antibacterial coverage to include agents effective against *Salmonella typhi* and other enteric pathogens may be appropriate. Similarly, empiric therapy for malaria, including for resistant infections due to *Plasmodium falciparum*, may be warranted.

MECHANICAL VENTILATION

Among patients with progressive respiratory failure, mechanical ventilation may be the best option. However, this modality is not typically available in resource-deprived areas. Techniques that minimize aerosolization of respiratory secretions are mandatory in order to prevent transmission to caregivers, as well as use of personal protective equipment that provides complete protection of skin and mucous membranes.

SUMMARY OF THE SUPPORTIVE MANAGEMENT OF SEVERE EBOLA VIRUS DISEASE

The reported mortality of EVD in prior outbreaks as well as the 2014–2016 West African outbreaks has ranged from 40% to 90%. Despite the knowledge accumulated during the 2014–2016 outbreaks about the most effective supportive care of patients with EVD including important information regarding management that emerged from the experience with imported cases in Europe and in the United States, insight into all aspects of care of the patient with severe disease is still incomplete. This is due in large part to the fact that almost all patients with EVD have been treated in resource-deprived settings where even data-gathering on the impact of specific treatment strategies is typically incomplete. Nonetheless, a variety of factors predictive of a fatal outcome have been identified in the recent and prior outbreaks, some of which may be important guides to the types of interventions likely to reduce the death rate. These poor prognostic factors include the following:

- Age greater than 45
- Respiratory, hemorrhagic, or neurologic manifestations
- Presenting complaints of weakness, dizziness, or diarrhea
- High levels of viral load
- Elevations of blood urea nitrogen, transaminases, creatinine, D-dimer, amylase, or nitric oxide
- Reduced number of T lymphocytes
- Impaired IgG and IgM response

(Dowell 1996, JID; Sanchez et al. 2004; Towner et al. 2004; Rollin et al. 2007; Schieffelin et al. 2014; WHO 2014; West and Von Saint Andre-von Amim 2014).
   West and Von Saint Andre-von Amim have summarized the key elements in management of severe EVD to include the following:

- Ensure personal safety and observe infection control protocols
- Assess for coinfection and superinfection
- Monitor and correct massive volume and electrolyte abnormalities using noninvasive monitoring techniques
- Use vasopressors if hypotension is not corrected by adequate fluid resuscitation and considerable adrenal insufficiency
- Administer blood products if necessary
- Manage acute kidney injury with fluid resuscitation and renal replacement therapy (see the section "Renal Replacement Therapy" on page 121) if indicated
- Manage respiratory failure optimally with mechanical ventilation using video laryngoscopy if required for intubation
- Consider intracranial hemorrhage and subclinical status epilepticus in obtunded patients
- Treat fever, pain, and anxiety as needed
- Provide nutrition by enteral route if possible
- Do not employ extracorporeal life support
- Consider experimental therapies
- Place central or peripheral intravenous catheter when repeated blood drawing is anticipated

- Perform continuous electrocardiographic and oxygen monitoring
- Use standard measures to prevent infection, stress ulceration, and venous thromboembolic complications
- Anticipate and address ethical issues around treatment
- Weigh risks and potential benefits of cardiopulmonary resuscitation
- Focus on comfort rather than on life-sustaining treatment in intractable disease

THERAPEUTIC AGENTS

Data from animal models and in vitro studies have yielded information helpful in the identification of promising therapies. None of these approaches has, as yet, been shown to be effective in randomized controlled trials of patients with EVD; however, data from animal models have been encouraging and there is a small amount of inconclusive clinical evidence derived from the treatment of individual patients in a compassionate use fashion.

In the years before the West African outbreak of 2014–2016, several therapeutic agents directed either at the virus itself or at key pathophysiologic processes involved in progression of infection were found to be effective to varying degrees in animal models. These included recombinant activated protein C, small molecules, vesicular stomatitis virus vaccine, and drugs directed at viral replication. Monoclonal antibody cocktails derived from humans and mice have demonstrated the greatest benefit, particularly when administered within 24 hours of infection in nonhuman primates (Qiu et al. 2014).

THERAPIES DIRECTED AT THE VIRUS

A number of agents directed at the virus have been studied in human and/ or animal models. None of these agents has yet been proven conclusively to be effective in therapy. The rapid decline in human cases of EVD beginning in early 2015 limited the opportunities for clinical trials. ZMapp, a combination of three monoclonal antibodies against the virus seemed to hold the most promise in a small number of patients, but in a clinical trial published in 2016, (Prevail II Writing Group 2016) it did not reach predetermined criteria for effectiveness (see below).

RIBAVIRIN

Ribavirin is an antiviral compound that interferes with the replication of several important RNA viruses. It is a prodrug that interferes with RNA-dependent nucleic acid synthesis by a mechanism that is not fully understood. Ribavirin has been used successfully in the treatment of hemorrhagic fevers caused by arenaviruses such as Lassa fever virus (Huggins 1989). It has proven ineffective, however, in blocking Ebola virus replication or protecting from EVD in animal models (Choi 2013).

ARTESUNATE–AMODIAQUINE

Empiric treatment for malaria was commonly given during the Ebola epidemic in West Africa in 2014–2016. During a 12-day period in the Ebola center in Foya, Liberia in August 2014, in which the first line agent artemether-lumefantrine was not available, the antimalarial combination drug artesunate–amodiaquine was substituted. During this period, a 31% reduction in mortality was observed among patients with Ebola virus infection (Gignoux et al. 2016). The reason for this was not clear and may have been the result of other factors.

ZMAPP

ZMapp is under development as a treatment for Ebola infection and was first used clinically in 2014 to treat a small number of patients during the West African epidemic. It has been studied in a randomized controlled trial at the time of this writing (2016). However, its safety and efficacy have not been fully established (see below). The drug is a blend of three chimeric (mouse–human) monoclonal antibodies directed at Ebola virus (Qiu et al. 2014). These neutralizing antibodies, which are manufactured in a tobacco plant, are designated c13C6, which is derived from a cocktail called MB-003 as well as two antibodies, c2G4 and c4G7 derived from the ZMab cocktail (Qiu 2014). MB-003 had previously been shown to be effective in Ebola virus infection in rhesus macaque monkeys (Olinger et al. 2012).

ZMapp was reported to provide complete protection of rhesus monkeys from fatal EVD (Qiu 2014) as much as five days after exposure and after the onset of severe symptoms.

Several patients were treated with ZMapp on a compassionate use during the West African outbreak (Hayden and Reardon 2014), although clinical effectiveness was difficult to confirm. Two of seven patients treated with ZMapp died (BMJ 2014).

A randomized, controlled trial comparing ZMapp (infusions given every third day for a total of three) to standard of care (Prevail II) was begun in March 2015 by the U.S. National Institute of Allergy and Infectious Diseases (NIAID) in partnership with the armed forces of Sierra Leone, Liberia, and Guinea. In this trial of 72 patients, 37% of patients receiving standard treatment and 22% of patients receiving standard therapy plus ZMapp died. Although the mortality rate was lower in patients receiving ZMapp, this result did not reach the statistical definition of efficacy (Prevail II Writing Group 2016).

BRINCIDOFOVIR

This drug is a prodrug of the antiviral agent cidofovir in which cidofovir is conjugated to a lipid matrix (Lanier et al. 2010; Quenelle et al. 2010). This allows entry into the host cell, where cidofovir is released in high concentration. Cidofovir is active in vitro against a variety of DNA viruses, including cytomegalovirus, adenovirus (Tollefson et al. 2014), variola (Olson et al. 2014), and vaccinia (Zaitseva et al. 2015). During the Ebola outbreak of 2014–2016, brincidofovir was one of a number of compounds tested for in vitro activity against Ebola virus in an urgent effort to identify potential therapies. Despite the fact that Ebola is an RNA virus, brincidofovir was found to limit its replication. Its mechanism for this effect is unclear. Several patients treated for EVD in the United States in 2014 were reported to have received brincidofovir. Its clinical efficacy could not be determined in this small number of cases. Because of declining numbers of EVD in West Africa in 2015, the manufacturer, Chimerix, Durham, North Carolina suspended clinical trials.

ALISPORIVIR

Alisporivir is a broad-spectrum antiviral compound that targets the host protein cyclophilin A (CypA). It is active against the flavivirus tick-borne encephalitis virus. However, only a modest or no effect has been demonstrated in tissue culture against several Ebola virus strains (Chiramel et al. 2016) because the virus does not depend on CryA for replication, unlike many other human viral pathogens.

TRIAZAVIRIN

Triazavirin is an antiviral compound with an azoloazine base structure that was developed in Russia as a treatment for influenza (Kaprenko et al. 2010). Because of its broad antiviral activity, it is being studied for possible use against Ebola (Darya 2014).

FAVIPIRAVIR (T-705, AVIGAN)

Favipiravir is an inhibitor of viral RNA polymerase (Furuta et al. 2013) that had demonstrated efficacy in animal models of Ebloa virus infection (Oestereich et al. 2014; Smither et al. 2014). However, no clinical efficacy or reduction in viral load was demonstrated in 99 adults and adolescents in several Ebola treatment centers (Sissoko et al. 2016).

TKM–EBOLA

Short-interfering RNA (siRNA), which binds to sequences in viral messenger RNA, has been demonstrated to block Ebola infection in animal models (Geisbert et al. 2006, 2010). A Phase I trial of siRNA (TKM–Ebola) began in January 2014 but was interrupted several months later because of possible side effects before adequate analysis of results was possible, and the drug was made available on a compassionate use protocol (Kraft et al. 2015). The study resumed in April 2015 using a lower dose of the drug. Development of the drug was suspended in July 2015 for apparent lack of efficacy.

BCX4430 (IMMUCILLIN-A)

This potential therapeutic agent, an adenosine analog with broad activity against a number of RNA viruses, was developed as potential treatment for hepatitis C. Activity against Ebola virus was demonstrated in animal models (Warren et al. 2014).

JK-05

A broad-spectrum antiviral agent developed in China and reported to have activity against a variety of RNA viruses including influenza and Ebola virus; it is thought to be similar in structure to favipiravir, described previously (Jourdan 2014).

FGI-106

A broad-spectrum antiviral agent active against several RNA viruses works by blocking viral entry into cells (Aman et al. 2009).

In a small number of patients infected with Ebola virus and managed in the United States, who received ZMapp, TKM–Ebola, brincidofovir, and/or convalescent plasma, no mutations were seen in the regions of the virus affected by these agents (Whitmer et al. 2016). Although this finding suggests that viral resistance is not necessarily a consequence of treatment, the small number of cases studied suggests that resistance mutations on the complete viral genome should continue to be sought.

NONSPECIFIC TREATMENTS

Therapies Directed at the Pathogenesis of Clinical Symptoms

Blocking the inflammatory manifestations of EVD

Infection with Ebola virus is characterized by the disruption of normal clotting pathways, particularly through the overproduction of tissue procoagulant factors. This results in disseminated intravascular coagulation (DIC) and the hemorrhagic complications associated with EVD. Among other effects, this process results in a reduction of circulating protein C (Geisbert et al. 2003a, 2003b). This phenomenon can result in the progression to multiorgan system failures accompanying progressive infection. This dysregulation of the clotting system has suggested potential targets for postexposure therapy of EVD in animal models. One approach has been to administer activated recombinant protein C in an effort to activate the protein C anticoagulant pathway (Hensley et al. 2007). Another strategy has been to block tissue procoagulant factor employing recombinant nematode anticoagulant protein C2 (Geisbert et al. 2003a). These interventions have been shown to extend survival in animal models but are of unknown utility in human infection.

Interruption of the clotting effects of EVD

Plasmapheresis

Because options for the treatment of Ebola virus disease remain limited, novel approaches to reducing or eliminating viremia continue to warrant interest. One such approach is extracorporeal plasmapheresis, a technique by which virus may be directly removed from the blood. Battner and colleagues reported the successful use of this technique in a patient with severe Ebola infection (Battner et al. 2014).

Convalescent serum

Historically, convalescent serum from patients who have recovered from EVD has been administered in a small number of cases (Emond et al. 1977). This approach was also used in the treatment of a small number of patients in the recent outbreak (WHO 2014) with no conclusions about efficacy yet reached. A study of the effectiveness of convalescent ZEBOV–Makona serum and *Sudan ebolavirus* (SEBOV) from macaques given to rhesus monkeys at the onset of viremia (three days after exposure) was only partially protective (Mire et al. 2016).

Therapies directed at the host response

As discussed in Chapter 7, infection with Ebola virus may result in a complex response of various elements of the host immune system. The *cytokine storm* seen in severe cases results in significant endothelial dysfunction, leading to profound fluid and electrolyte disorders. Medications known to support and potentially restore endothelial function were administered to a number of patients in the 2014–2016 epidemic. Statins such as simvastatin (Chen et al. 2007), and atorvastatin (Xiao et al. 2013), medications used to treat hypercholesterolemia, are known to protect the endothelial barrier. The angiotensin receptor blockers (ARBs) may also restore endothelial integrity (Bodor et al. 2012). These agents, used singly and in combination, may improve survival in some patients with pneumonia (Mortensen et al. 2012), and, perhaps, severe sepsis from other causes. In addition, statins have been shown to be inhibitory toward several RNA viruses (Fedson et al. 2015).

Combination therapy with atorvastatin (40 mg daily) and the ARB agent irbesartan (150 mg daily) was administered to approximately 100 individuals in Liberia (Fedson et al. 2015) in an open-label study. Although circumstances made complete data gathering difficult, according to the authors, rapid clinical improvement was described in all patients. This observation has not yet been confirmed in randomized trials. If proven effective and safe, therapy with statins and/or ARBs could represent an inexpensive and effective strategy. The potential agents, which are generic and are in widespread use, could potentially be made available in resource-deprived areas prior to the development of definitive therapy for EVD (Fedson and Rordam 2015). This so-called *bottom-up* approach—that is, a strategy in which the effects of the virus are targeted rather than the virus itself—could potentially be used in combination with the *top-down* approach, emphasizing antivirals.

REFERENCES

Aman MJ, Kinch MS, Warfield K et al. Development of a broad-spectrum antiviral with activity against Ebola virus. *Antiviral Res* 2009;83(3):245–251.

Bah EI, Lamah M, Fletcher T et al. Clinical presentation of patients with Ebola virus disease in Conakry, Guinea. *N Engl J Med* 2015;372:40–47.

Battner S, Koch B, Delnik O et al. Extracorporeal virus elimination for the treatment of severe Ebola virus disease—First experience with lectin affinity plasmapheresis. *Blood Purif* 2014;38(3–4):2867–2891.

BMJ. US signs contract with ZMapp maker to accelerate development of the Ebola drug. *BMJ* 2014;349:g5488.

Bodor C, Nagy JP, Vegh B et al. 2012. Angiotensin II increases the permeability and PV-1 expression of endothelial cells. *Am J Physiol Cell Physiol* 2011;302:C267–C276. doi:10.1152/ajpcell.00138.2011.

Chen W, Pendyala S, Natarajan V, Garcia JG, Jacobson JR. Endothelial cell barrier protection by simvastatin: GTPase regulation and NADPH oxidase inhibition. *Am J Physiol Lung Cell Mol Physiol* 2007;295:L575–L583. doi:10.1152/ajplung.00428.2007.

Chertow DS, Uyeki TM, DuPont HL. Loperamide therapy for voluminous diarrhea in Ebola virus disease. *J Infect Dis* 2015;211:1036.

Chiramel AI, Banadyga L, Dougherty JB et al. Alisporivir has limited antiviral effects against Ebola virus strains Makona and Mayinga. *J Infect Dis* 2016;214(S3):S355–S359.

Choi JH, Croyle MA. Emerging targets an novel approaches to Ebola virus prophylaxis and treatment. *BioDrugs* 2013;27:565–583.

Darya K. New antiviral drug from Urals will help fight Ebola and other viruses. *Russia Beyond the Headlines*, November 12, 2014.

Dowell SF. Ebola hemorrhagic fever: Why were children spared? *Padiatr Infect Dis J* 1996;15(3):189–191.

Emond RT, Evans B, Bowen ET, Lloyd G. A case of Ebola virus infection. *BMJ* 1977;2(6086):541–544.

Fedson DS, Jacobson JR, Rordam OM, Opal S. Treating the host response to Ebola virus disease with generic statins and angiotensin receptor blockers. *mBio* 2015;6(3):e00716-15. doi:10.1128/mBio.00716-15

Fedson DS, Rordam OM. Treating Ebola patients: A "bottom up" approach using generic statins and angiotensin receptor blockers. *Int J Infect Dis* 2015. http://creativecommons.org/licenses/by-nc-nc/4.0.

Furuta Y, Gowen BB, Takahashi K et al. Favipiravir (T-705), a novel viral RNA polymerase inhibitor. *Antiviral Res* 2013;100:446.

Geisbert TW, Young HA, Jahrling PB et al. Mechanisms underlying coagulation abnormalities in Ebola hemorrhagic fever: Overexpression of tissue factor in primate monocyte/macrophages is a key event. *J Infect Dis* 2003a;188(11):1618–1629.

Geisbert TW, Hensley LE, Jahrling PB et al. Treatment of Ebola virus infection with a recombinant inhibitor of factor VIIa/tissue factor: A study in rhesus monkeys. *Lancet* 2003b;362(9400):1953–1958.

Geisbert TW, Hensley LE, Kagan E et al. Postexposure protection of guinea pigs against a lethal ebola virus challenge is conferred by RNA interference. *J Infect Dis* 2006;193:1650.

Geisbert TW, Lee AC, Robbins M et al. Postexposure protectionof non-human primates against a lethal Ebola virus challenge with RNA interference: A proof-of-concept study. *Lancet* 2010;375:1896.

Gignoux E, Azman AS, De Smet M et al. Effect of artesunate-amodiaquine on mortality related to Ebola virus disease. *N Engl J Med* 2016;374:23–32.

Hayden EC, Reardon S. Should experimental drugs be used in the Ebola outbreak? *Nature*, August 12, 2014.

Hensley LE, Stevens EL, Yan SB et al. Recombinant human activated protein C for the postexposure treatment of Ebola hemorrhagic fever. *J Infect Dis* 2007;196(Suppl 2):S390–S399.

Huggins JW. Prospects for treatment of hemorrhagic fevers with ribavirin, a broad-spectrum antiviral drug. *Rev Infect Dis* 1989;11(Suppl 4):S750–S761.

Jourdan A. China military-linked firm eyes quick approval of drug to cure Ebola. *Yahoo News*, October 14, 2014.

Kaprenko I, Deev S, Kiselev O et al. Antiviral properties, metabolism, and pharmacokinetics of a novel azolo-1,2,4-triazine-derived inhibitor of influenze A and B virus replication. *Antimicrob Agents Chemother* 2010;54(5):2017–2022.

Kraft CS, Hewlett AL, Koepsell S et al. The use of TKM-100802 and convalescent plasma in 2 patients with Ebola virus disease in the United States. *Clin Infect Dis* 2015;61:496.

Lanier R, Trost L, Tippin T et al. Development of CMX001 for the treatment of poxviruses infections. *Viruses* 2010;2(12):2740–2762.

Mire CE, Geisbert JB, Agans KN et al. Passive immunotherapy: Assessment of convalescent serum against Ebola virus Makona infection in nonhuman primate. *J Infect Dis* 2016;214(S3):S367–S374.

Mortensen EM, Nakashima B, Cornell J et al. Population-based study of statins, angiotensin II receptor blockers and angiotensin-converting enzyme inhibitors on pneumonia-related outcomes. *Clin Infect Dis* 2012;55:1466–1473.

Oestereich L, Ludtke A, Wurr S et al. Successful treatment of advanced Ebola virus infection with T-705 (favipiravir) in a small animal model. *Antiviral Res* 2014;105:17.

Olinger CG, Pettitt J, Kim D et al. Delayed treatment of Ebola virus infection with plant-derived monoclonal antibodies provides protection in rhesus macaques. *Proc Natl Acad Sci* 2012;109(44):18030–18035.

Olson VA, Smith SK, Foster S et al. In vitro efficacy of brincidofovir against variola virus. *Antimicrob Agents Chemother* 2014;58:5570.

Prevail II Writing Group. A randomized, controlled trial of ZMapp for Ebola virus infection. *N Engl J Med* 2016;375:1448–1456.

Qiu X, Wong G, Audet J et al. Reversion of advanced Ebola virus disease in nonhuman primates with ZMapp. *Nature* 2014;514:47–54.

Quenelle DC, Lampert B, Collins DJ et al. Efficant of CMX001 against herpes simplex virus infections in mice and correlations with drug distribution studies. *J Infect Dis* 2010;202(10):1492–1499.

Rollin PE, Bausch DG, Sanchez A. Blood chemistry measurements and D-Dimer levels associated with fatal and nonfatal outcomes in humans infected with Sudan Ebola virus. *J Infect Dis* 2007;196(Suppl 2):5364–5371.

Sanchez A, Lukwiya M, Bausch D et al. Analysis of human peripheral blood samples from and nonfatal cases of Ebola (Sudan) hemorrhagic fever: Cellular responses, virus load and nitric oxide levels. *J Virol* 2004;78:10370–10377.

Schieffelin JS, Shaffer JG, Goba A et al. Clinical illness and outcomes in patients with Ebola in Sierra Leone. *N Engl J Med* 2014;371(22):2092–2100.

Sissoko D, Laouenan C, Folkesson E et al. Experimental treatment with Favipiravir for Ebola virus disease (the JIKI trial): A historically controlled single-arm, proof-of-concept trial in Guinea. *PLoS Med* 2016;13:e1001967.

Smither SJ, Eastaugh LS, Steward JA et al. Post-exposure efficacy of oral T-705 (Favipiravir) against inhalational Ebola virus infection in a mouse model. *Anitviral Res* 2014;104:153.

Tollefson AE, Spencer JF, Ying B et al. Cidofovir and brincidofovir reduce the pathology caused by systemic infection with human type 5 adenovirus in immunosuppressed Syrian hamsters, while ribavirin ins largely ineffective in this model. *Antiviral Res* 2014;112:38–46.

Towner JS, Rollin PE, Bausch DG et al. Rapid diagnosis of Ebola hemorrhagic fever by reverse transcription-PCR in an outbreak setting and assessment of patient viral load as a predictor of outcome. *J Virol* 2004;78:4330–4341.

Warren TK, Wells J, Panchal RG et al. Protection against filovirus diseases by a novel broad-spectrum nucleoside analogue BC4430. *Nature* 2014;508(7496):402.

West TE, Von Saint Andre-von Amim A. Clinical presentation and management of severe ebola virus disease. *Ann Am Thorac Soc* 2014;11(9):1341–1350.

Whitmer SLM, Albarino C, Shepard SS et al. Preliminary evaluation of the effect of investigational Ebola virus disease treatments on viral genome sequences. *J Infect Dis* 2016;214(S3):S333–S341.

Wolf T, Ross MJ, Davenport A. Minimizing the risks associated with renal replacement therapy in patients with Ebola virus disease. *Kidney Int* 2015;87:5–7.

World Health Organization (WHO). Use of convalescent whole blood or plasma collected from patients recovered from Ebola virus disease for transfusion, as an empirical treatment during outbreaks. http://apps.who.int/iris/bitstream/10665/135591/1/WHO_HIS_SDS_2014_eng.pdf.

Xiao H, Qin X, Ping D, Zuo K. Inhibition of Rho and Rac geranylgeranylation by atorvastatin is critical for preservation of endothelial junction integrity. *PLoS One* 2013;8:e59233. doi:10.1371/journal.pone.0059233.

Zaitseva M, McCullough KT, Cruz S et al. Postchallenge administration of brincido-fovir protects healthy and immune-deficient mice reconstituted with limited numbers of T cells from lethal challenge with IHD-J-Luc vaccine virus. *J Virol* 2015;89(6):3295–3307.

Vaccine Development

9

At the onset of the 2014–2016 West African outbreak of Ebola virus disease (EVD), no vaccine to prevent the disease had yet undergone human trials. Less than two years later, a candidate vaccine had been tested in West Africa, and results pointed to significant effectiveness against the *Zaire ebolavirus* (ZEBOV) (Henao-Restrepo et al. 2016). The durability of protection conferred by this vaccine is not yet known, and its effectiveness against other strains of Ebola remains to be seen. Other candidate vaccines also are at various stages of testing. Yet, the implications of this favorable finding are dramatic in terms of potentially preventing future widespread outbreaks of EVD.

BRIEF HISTORY OF VACCINE EFFORTS

This apparently quick progress in vaccine development had roots reaching back several decades. Efforts to develop a vaccine to prevent Ebola and the related Marburg virus began in the 1980s, not long after Ebola was first identified in humans in 1976 (Marzi and Feldmann 2014). However, with Ebola occurring in relatively small clusters in remote rural areas of East and Central Africa, interest in the development of a vaccine was somewhat limited. Interest was motivated mainly by the extremely high mortality rate and by the need to protect not only the public in these areas but, importantly, health care workers who would be involved in treatment and containment efforts. Work to develop a vaccine against Ebola and Marburg was regarded with increased urgency because of concern about bioterrorism after the September 11, 2001 attacks on the World Trade Center and the Pentagon and the subsequent anthrax attacks. In these biological attacks, anthrax spores were sent through the U.S. Mail, killing five people and causing major dislocation to the U.S. Congress, the U.S. Postal Service, and public health departments around the country (Levine et al. 2015). The U.S. Centers for Disease Control and Prevention (CDC) considered Ebola, Marburg, and other viruses as Category A bioterrorism agents, that is among the likeliest agents to be used in a potential biologic attack. (See Chapter 10 for a more detailed discussion of Ebola's potential as a bioweapon.)

In 2014, as the West Africa Ebola outbreak spread, vaccine development became a much more immediate priority. Although several candidate vaccines had previously shown promise in animal and in vitro studies, none had been established as safe and effective in humans. Vaccines had drastically reduced the incidence of a number of important viral and bacterial diseases worldwide, and there was legitimate hope for an Ebola vaccine to ultimately be developed. Yet, the urgency of the circumstances in West Africa and the extremely high level of fear and anxiety about EVD in the rest of the world tended to lead to unrealistic expectations for the speed at which a vaccine could be developed.

TARGETS FOR VACCINE DEVELOPMENT

Ebola and Marburg virus are both negative-strand RNA viruses and members of the Filoviridae family. As discussed in Chapter 1, Ebola virus (EBOV) is separated into five species:

- *Zaire ebolavirus* (ZEBOV)
- *Sudan ebolavirus* (SEBOV)
- *Bundibugyo ebolavirus* (BEBOV)
- *Tai Forest ebolavirus*
- *Reston ebolavirus*

Both ZEBOV and SEBOV have been associated with human outbreaks with a mortality rate of 40%–90%, including the West African outbreak of 2014–2016 caused by ZEBOV. These species have been targeted for vaccine development.

Ebolavirus preferentially targets cells of the immune system, including macrophages and dendritic cells. After viral attachment and infection of these cells, they release glycoproteins of viral origin. In this way, the virus can deregulate the dendritic cell response and interferes with their maturation and the development of an adaptive immune response. Infected macrophages overexpress proinflammatory cytokines which, if unchecked, may result in the *cytokine storm* through which infection causes capillary leak and end-organ damage.

ANIMAL MODELS FOR VACCINE TRIALS

Two primates, the rhesus and the cynomolgus monkey, had been established as the usual animal models for research into Ebola virus vaccines. Another primate, the macaque monkey, when infected develops a disease most similar to Ebola virus disease (EVD) seen in humans. A number of rodent models also have been used in vaccine trials to establish immunogenicity of various vaccine preparations. Because of the difficulty in conducting human trials of vaccines against sporadic and highly lethal conditions like EVD, the United States Food and Drug Administration (FDA) has asserted the so-called animal rule for the development and approval of such vaccines. This rule permits licensing of vaccines if they have been demonstrated to be safe and effective in appropriate animal models. It was applied by the FDA for the first time in 2015, when it licensed the anthrax vaccine BioThrax (Beasley et al. 2016).

CANDIDATE VACCINES AND RESULTS OF TRIALS TO DATE

The earliest potential vaccines, developed shortly after Ebola virus was first identified in 1976, were composed of inactivated ZEBOV virus. They were found to be effective in rodents but not in nonhuman primates (Lupron et al. 1980). Following these attempts, subunit vaccines consisting of recombinant portions of ZEBOV were studied but were found to have insufficient immunogenicity in rodents (Hoenen et al. 2012). Later efforts focused on the use of protein-based complexes consisting of various combinations of ZEBOV matrix protein (VP40) and glycoprotein (GP) and nucleoprotein (NP). GP, which aids in viral entry into phagocytic cells, endothelial cells, and hepatocytes, is the primary antigen targeted by current candidate vaccines (Sarwar et al. 2015). Some of these later efforts produced more robust immune responses in rodents and protection in nonhuman primates (Warfield et al. 2007). Vaccines employing DNA plasmids and vectors such as vesicular stomatitis and recombinant adenovirus have been studied in more recent years (Hart 2003; Jones et al. 2005; Martin et al. 2006; Daddario-DiCaprio et al. 2006a, 2006b; Sarwar et al. 2011).

One candidate vaccine—an adenovirus vaccine expressing an Ebola variant Makona glycoprotein—was found to be 100% protective in guinea pigs against

Ebola virus adapted to the animals when they were challenged four weeks after vaccination (Wu et al. 2016).

Another candidate vaccine, a DNA vaccine, underwent a Phase I clinical trial in 20 human volunteers and demonstrated significant immunogenicity and no significant adverse reactions (Sarwar et al. 2015).

Because of the urgency of the West African outbreak, several fast track vaccine trials in humans were initiated in West Africa in 2015.

Trial of rVSV Backbone Vaccine, Guinea

The most impressive results reported so far were in Phase I–III studies of a vaccine employing the recombinant livestock virus vesicular stomatitis virus (rVSV) backbone, which had been shown to be effective in animal models. The vaccine was produced by replacing the section of VZV RNA that codes for surface protein with the section of *ebolavirus*, which codes for the surface glycoprotein and using this altered rVSV, termed rVSV–ZEBOV, as the vaccine preparation. It was originally produced several years before the West African outbreak by the United States Army and the Public Health Agency of Canada and is licensed to the Merck Corporation. In a trial in Guinea led by the WHO, the Guinean Health Ministry, the Norwegian Institute of Public Health, and a variety of other organizations, the vaccine demonstrated a 100% rate of protection 10 days or more after vaccination. Full results of the rVSV–EBOV vaccine trial were published in late 2016 (Henao-Restrepo et al. 2016). The study was completed after the West African epidemic had been largely brought under control in Basse–Guinee, a region of Guinea still seeing cases by early 2015. The subjects were 11,841 residents of Guinea. A so-called *ring vaccination* strategy was used in which persons who had contact with a subject with confirmed Ebola virus disease within 21 days were randomized to receive the vaccine or not. Each such ring included approximately 80 individuals, such as family members, neighbors, and caregivers. Persons within the contact group who developed EVD within the first nine days after vaccination were assumed to have been infected before receiving the vaccine and were not included in the analysis. Using this form of vaccination strategy and this analysis, there were no cases of EVD among the individuals vaccinated (5837) and 23 cases among those not receiving the vaccine. Headache (25.4%), fatigue (18.9%), and myalgia (13.1%) were the most commonly reported side effects of the vaccine. Of 80 serious adverse events, three were considered to be possibly related to the vaccine, one febrile reaction, one influenza-like illness, and one case of anaphylaxis. Each patient recovered from these uneventfully.

Largely on the basis of the preliminary results of this study, an emergency supply of this vaccine of 300,000 was planned and funded through an agreement between the Merck Corporation and Gavi, the Vaccine Alliance, (Westcott 2016), which includes the WHO, UNICEF, the Bill and Melinda Gates Foundation, and the World Bank.

Although this vaccine was the first demonstrated to prevent infection from the West African or ZEBOV strain of Ebola virus in humans, the duration of protection provided by this vaccine and its effectiveness against the other strains of the virus are not yet known.

Trial of rRSV Backbone Vaccine, Sierra Leone

A trial in Sierra Leone called STRIVE (Sierra Leone Trial to Introduce a Vaccine against Ebola) conducted by the College of Medicine and Allied Health Sciences of the University of Sierra Leone, the Sierra Leone Ministry of Health and Sanitation, and the U.S. Centers for Disease Control also employed rVSV–ZEBOV and focused on health care and other frontline workers. It recently entered Phase III, with 8673 participants enrolled in this trial, including 453 and 539 in safety and immunogenicity studies (Widdowson et al. 2016) between April and August 2015. Although no Ebola cases were seen in the study group, no conclusion regarding vaccine efficacy could be reached because of the low overall case frequency during that period. However, no serious vaccine-related adverse events were encountered.

Other vaccine trials that were in earlier stages of human study at the time of this writing:

- Chimp adenovirus 3 vectored glycoprotein (Glaxo-SmithKline, National Institute of Allergy and Infectious Diseases) (Pavot 2016) entered Phase III clinical trial in 2016
- Human adenovirus 5 vectored 2014 glycoprotein insert (BIT and CanSino): Phase I complete (Zhu et al. 2015)
- Adenovirus 26 vectored glycoprotein/MVA-BN (Ad26.ZEBOV/MVA-BN) (Johnson & Johnson): Phase I complete (Milligan et al. 2016)

Because the Ebola virus has five known subtypes and is related to the Marburg virus, which also causes severe disease and has a high mortality rate, an ideal vaccine might be effective against all known strains, as well as Marburg virus. However, the feasibility of using the VSV spine to create EVD vaccines against multiple strains might be limited by side effects. The results of trials of other vaccines under development at this time of this writing may follow.

CLINICAL AND ETHICAL CHALLENGES OF EBOLA VACCINE TRIALS

Ethical Issues

Convincing data about the safety and efficacy of vaccines most often comes from clinical trials. The need for a placebo group in any randomized clinical trial raises concerns about the ethics of depriving the placebo group of an effective vaccine and allowing them to have a higher risk of contracting the infection. This quandary is not unique to potential Ebola vaccines. Two ethical principles often are cited as obstacles to conducting randomized controlled trials in the developing world. These are the ethical principle of beneficence, the need to be as certain as possible that any clinical trial is designed to do good for the population being studied, and the principle of equipoise, which requires that all participants have an equal likelihood of being benefited or harmed by a trial. Equipoise, in particular, which implies that there is genuine uncertainty about the benefit of the treatment arm in a controlled trial, may be difficult to maintain in a study of a vaccine intended to prevent a highly lethal infection like EVD in a community where infection is rapidly spreading.

The potential challenges in adhering to ethical standards have arisen in relation to trials of other vaccines studied in Africa.

Clinical Issues

The field testing of vaccines in an outbreak setting raises substantial challenges (Bausch et al. 2008).

Among these would likely be:

- *Developing the means to manufacture, transport, and store vaccine*: Vaccine production would of necessity be carried out in sophisticated laboratory facilities, not necessarily in regions most impacted by a rapidly spreading epidemic. Storage, which probably would require refrigeration and careful monitoring of expiration times, may pose challenges to rapid deployment.
- *Developing systems of recording vaccine administration*: Accurate and consistent means of recording the time and date of vaccination of identifiable individuals would be essential, as would the

development of a system of follow-up and contact to evaluate for both efficacy and safety of the vaccine.

- *Confirming accurate and timely diagnosis*: Other illnesses, particularly malaria and typhoid, may be much more common than the target illness, even during an outbreak of the target illness. This was the case during the Ebola outbreak in West Africa in 2014–2016. This fact requires a system of accurate and rapid diagnosis of the illness targeted by the vaccine under evaluation as well as other common diseases with similar symptoms.

- *Providing timely vaccination to potential victims prior to their becoming infected*: Identification of persons at risk of contagion can represent a significant challenge, especially in a rapidly expanding epidemic. In addition to household contacts, employment and casual contacts, among others, would have to be rapidly and accurately identified in order to provide vaccination before infection. This challenge would be even more substantial with an infection that, unlike Ebola, spreads by airborne or foodborne routes.

- *Monitoring, categorizing, and managing vaccine side effects*: The majority of vaccine side effects probably would not be immediate. Monitoring for late side effects such as fever, respiratory difficulties, or diarrhea, as well as potential abnormalities such as hepatitis, renal insufficiency, cytopenias, and others, would require a workable plan for follow-up of vaccine recipients.

- *Following vaccinated individuals to assess effectiveness, both immediate and late*: Individuals who have been vaccinated may have been infected subclinically before receiving the vaccine. In order to accurately determine vaccine effectiveness, laboratory proof that a vaccine recipient had not already been infected would be optimal. Lacking this, a prolonged incubation period in a person who was vaccinated despite having asymptomatic infection could be mistakenly interpreted as a protective response to the vaccine. For this reason, late follow-up may be necessary to establish vaccine efficacy accurately.

These critical issues would become more difficult to measure and analyze in the evaluation of a vaccine to prevent a disease that is rapidly spreading and that poses a grave threat to caregivers. In addition, in view of the logistical difficulties in immunizing large numbers of people in an outbreak setting, the potential feasibility of nasal aerosol vaccination effective in animal models, has been proposed (Jonsson-Schmunk and Croyle 2015) for a possible answer to cost and compliance issues.

REFERENCES

Bausch DG, Sprecher A, Jeffs B, Boumandouki P. Treatment of Marburg and Ebola hemorrhagic fevers: A strategy for testing new drugs and vaccines under outbreak conditions. *Antiviral Res* 2008;78:150–161.

Beasley DW, Brasel TL, Comer JE. First vaccine approval under the FDA animal rule. *Npj Vaccines* 2016;1:16013. doi:10.1038/npjvaccines.2016.13; published online August 25, 2016.

Daddario-DiCaprio KM, Geisbert TW, Geisbert JB et al. Cross-protection against Marburg virus strains by using a live, attenuated recombinant vaccine. *J Virol* 2006a;80:9659–9666.

Daddario-DiCaprio KM, Geisbert TW, Stroher U et al. Postexposure protection Marburg heaemorrhagic fever with recombinant vesicular stomatitis virus vectors in non-human primates: An efficacy assessment. *Lancet* 2006b;367:1399–1404.

Hart MK. Vaccine research efforts for filoviruses. *Int J Parasitol* 2003;33:583–595.

Henao-Restrepo AM, Camacho A, Longini IM et al. Efficacy and effectiveness of an rVSV-vectored vaccine in preventing Ebola virus disease: Final results from the Guinea ring vaccination, open-label, cluster-randomised trial (Ebola Ca Suffit!). *Lancet* 2017;389:505–518. doi:10.1016/S0140-6736(16)32621-6.

Hoenen T, Groseth A, Feldmann H. Current Ebola vaccines. *Expert Opin Biol Ther* 2012;12:859–872.

Jones SM, Feldmann H, Stroher U et al. Live attenuated recombinant vaccine protects nonhuman primates against Ebola and Marburg viruses. *Nat Med* 2005;11:786–790.

Jonsson-Schmunk K, Croyle MA. A long-lasting, single dose nasal vaccine for Ebola: A practical armament for an outbreak with significant global impact. *Expert Rev Anti Infect Ther* 2015;13(5):527–530. doi:10.1586/14787210.2015.1028368.

Levine MM, Tapia M, Hill AV, Sow SO. How the current West African Ebola virus disease epidemic is altering views on the need for vaccines and galvanizing a global effort to field-test leading candidate vaccines. *J Infect Dis* 2015;211:504–507.

Lupron HW, Lambert RD, Bumgardner DL et al. Inactivated vaccine for Ebola virus efficacious in guinea pig mode. *Lancet* 1980;2:1294–1295.

Martin JE, Sullivan NJ, Enama ME et al. A DNA vaccine for Ebola virus is safe and immunogenic in a phase I clinical trial. *Clin Vaccine Immunol* 2006;13:1267–1277.

Marzi A, Feldmann H. Ebola virus vaccines: An overview of current approaches. *Expert Rev Vaccines* 2014;13(4):521–531.

Milligan ID, Gibani MM, Sewell BA et al. Safety and immunogenicity of novel adenovirus type 26- and modified vaccinia Ankara-vectored Ebola vaccines. A randomized clinical trial. *JAMA* 2016;315(15):1610–1623. doi:10.1001/jama.2016.4218.

Pavot V. Ebola virus vaccines: Where do we stand? *Clin Immunol* 2016;173:44–49.

Sarwar UN, Costner P, Enama ME et al. Safety and immunogenicity of DNA vaccines encoding Ebolavirus and Marburgvirus wild-type glycoproteins in a phase I clinical trial. *J Infect Dis* 2015;211:549–557.

Sarwar UN, Sitar S, Ledgerwood JE. Filovirus emergence and vaccine development: A perspective for health care practitioners in travel medicine. *Travel Med Infect Dis* 2011;9:126–134.

Warfield KL, Swenson DL, Olinger GG et al. Ebola virus-like particle-based vaccine protects nonhuman primates against lethal Ebola virus challenge. *J Infect Dis* 2007;196(Suppl 2):S430–S437.

Westcott L. $5 Million Ebola vaccine deal announced at Davos. *Newsweek*, January 20, 2016. http://www.newsweek.com/ebola-vaccine-merck-gavi-5-million-417774.

Widdowson M, Schrag SJ, Carter RJ et al. Implementing an Ebola vaccine study— Sierra Leone. *MMWR* 2016;65(3):98–106.

Wu S, Kroeker A, Wong G et al. An adenovirus vaccine expressing Ebola virus variant Makonaglyhcoprotein is efficacious in Guinea pigs and nonhuman primates. *J Infect Dis* 2016;214(S3):S326–S332.

Zhu F, Hou L, Li J et al. Safety and immunogenicity of a novel recombinant adenovirus type-5 vector-based Ebola vaccine in healthy adults in China: Preliminary report of a randomized, double-blind, placebo, phase I trial. *Lancet* 2015;385(9984):2272–2279. doi:10.1016/S0140-6736(15)60553-0.

PART THREE

Tools for Preparing

Potential Bioterrorism Concerns

10

Since the terrorist attacks on the United States on September 11, 2001, there has been an increased focus on the possibility of bioterrorism (Masci and Bass 2005), potentially by both governmental and nongovernmental perpetrators. This concern typically has been raised about two bacteria, *Bacillus anthracis* and *Yersinia pestis*, the agents of anthrax and plague, respectively, and variola virus, the cause of smallpox. As shown in Table 10.1, these organisms top the so-called Category A list developed by the United States Centers for Disease Control and Prevention (CDC). They were assigned this status for several reasons, including presumed ease of dispersal and extremely high mortality rates, which give them the potential to cause mass casualties. Also on this high-probability list are the so-called hemorrhagic fever viruses, including Ebola virus, as well as dengue, Lassa fever, Marburg, and several other viral agents (Cunha 2002; Kagan 2005; Cenciarelli et al. 2015; Passi et al. 2015).

If used in a biological attack, Ebola virus would require direct person-to-person physical contact, unlike smallpox, plague, or anthrax. Both smallpox and inhalational anthrax are transmitted primarily through the respiratory tract by aerosol droplets. Plague, in its inhalational form, is transmitted in a similar fashion and is also transmissible by the bite of an infected flea. Both smallpox and plague carry the risk of person-to-person transmission. Transmission by the respiratory tract has not been documented with Ebola virus. Nonetheless, even limited to direct person-to-person transmission, Ebola virus would pose the risk of great lethality, and challenges in providing medical resources to treat large numbers of patients while protecting health care workers from infection. Any of these putative agents of bioterrorism, if released, would cause great public anxiety and likely panic, with the attendant overwhelming demand placed on health care services and emergency response systems.

TABLE 10.1   Selected Category A potential bioterrorism agents. This chart shows selected disease agents on the U.S. Centers for Disease Control and Prevention's Category A list of potential bioterrorism agents. Category A contains the agents deemed to pose the greatest risk from possible bioterrorism. Ebola is included as part of a broader category of hemorrhagic fever viruses, which also includes dengue, Lassa fever, Marburg, Hantaviruses, and others. Also included in Category A are botulism (*Clostridium botulinum*) and tularemia (*Francisella tularensis*). Category A agents can be easily spread or transmitted from person to person, have high mortality rates, might cause panic, and would require special preparedness from health care institutions

	CAUSE	MODES OF TRANSMISSION	INCUBATION PERIOD	MORTALITY RATE (CASE FATALITY RATE)	TREATMENT	PREVENTION
Smallpox	*Variola virus*	Airborne droplets from infected person; fomites	10–12 days	33% (untreated)	Vaccine given within 3 days of exposure	Effective vaccine exists
Anthrax	*Bacillus anthracis*	Inhalational (inhaling spores) Cutaneous (contact with soil, animal hides) Intestinal (eating infected meat)	5–11 days (inhalational form)	45%–85% (treated U.S. cases in the twentieth and twenty-first centuries)[a]	Combination antibiotic therapy for 60 days	Vaccine requires several doses over 18 months with annual booster injections
Plague	*Yersinia pestis*	Airborne droplets (pneumonic form) Bite of infected fleas or rodents (bubonic form)	1–4 days or 1–6 days	About 66% (untreated) 11% (treated in U.S.)[c]	Antibiotic therapy begun within 24 hours, continued for 7 days	Old vaccine not available in U.S. New vaccine in development[b]
Ebola	*Ebolavirus*	Direct person-to-person contact, possibly animal contact	1–21 days	20%–90%	No proven therapy (aside from supportive case and fluid replacement)	Vaccine in development

Source (except where noted): Masci, J.R. and Bass, E., *Bioterrorism: A Guide for Hospital Preparedness*, CRC Press, Boca Raton, FL, 361pp, 2005.

[a] "Anthrax" National Center for Emerging and Zoonotic Infectious Diseases, CDC, 2009, https://www.cdc.gov/nczved/divisions/dfbmd/diseases/anthrax/technical.html.

[b] Plague: Prevention, CDC, 2015, https://www.cdc.gov/plague/prevention/index.html.

[c] Plague: Frequently asked questions, CDC, 2015, https://www.cdc.gov/plague/faq/.

Key elements of potential effectiveness of biological agents as weapons of terrorism are the potential for causing widespread fear beyond, perhaps, the actual risk posed—as well as disruption of the normal functioning of political structures and health care systems. Based on the global reaction to the Ebola outbreak of 2014–2016, the virus would be potentially effective as an agent of terrorism. The events of this outbreak stimulated further discussion and analysis of the possible use of Ebola virus in this way (Gunaratne 2015).

What follows is a discussion of the key factors to be considered in establishing the likelihood of Ebola virus being used in an intentional attack, as well as the potential attack scenarios and the impact on the health care system. Each of the potential agents of biological attack and bioterrorism would pose important challenges to the health care system. Outside of a known outbreak or attack setting, a delay in recognition of the clinical syndromes associated with the likely pathogens would seem to be inevitable. In the case of an intentional release of Ebola virus, the need for the rapid deployment of effective infection control strategies and for equipping and training medical staff would be truly challenging. Nonetheless, obtaining the virus and working with it for such a purpose would also be extremely challenging.

SPECIFIC CHALLENGES POSED BY EBOLA VIRUS

As was seen in the 2014–2016 West Africa outbreak of Ebola virus disease (EVD), once Ebola is introduced into a population, it has high potential for person-to-person spread, including to health care workers and first responders. Although this form of transmission requires direct contact with body fluid and would be unlikely to result in large numbers of victims in developed countries, the potential for secondary infection of health care workers is significant, as was seen in West Africa and in the United States. Even in the United States and other developed countries in which patients with EVD were treated during that outbreak, health care facilities came under great pressure to provide effective means of screening patients for potential exposure to EVD, to provide appropriate isolation and quarantine logistics, and to maintain an environment in which first responders and health care workers could effectively transport, evaluate, and treat potential victims. In those developed countries, public fears of contagion resulted in high levels of anxiety, as well as implementation of confusing and often contradictory public health measures.

HOW AN ATTACK MIGHT
BE CARRIED OUT

If a biological attack using Ebola virus were to be carried out, several important obstacles would have to be overcome by the perpetrators.

- In its current form, Ebola virus cannot be effectively transmitted by the aerosol route and does not survive long on inanimate objects. The organism would have to be obtained in a form in which it could be used to infect individuals by injection or by direct contact with body fluid or another liquid containing the virus. In the absence of human cases or an animal vector, this would likely mean access to high-level containment facilities where the virus is being isolated or used in vaccine or drug development. Obviously, working with the virus outside of appropriate containment would pose a high risk to the perpetrators, a fact that might render Ebola virus and similar agents impractical for this purpose. Transmission by infected animals, such as bats, however, is conceivable. In addition, as has been pointed out, the risk to perpetrators, although high, might be considered acceptable (Gunaratne 2015).
- A strategy would have to be developed to introduce the virus into a population despite the above constraints in transmission. It has been pointed out that Ebola virus could be employed as an agent of bioterrorism in a variety of settings (Cenciarelli et al. 2015). Among these potentially are travel hubs including airports, cruise ships, and subway stations.

One feared scenario common to several of the potential agents of bioterrorism is the introduction to a closed population by means of a suicidal volunteer.

Transmission by direct personal contact during the early, relatively asymptomatic phase of Ebola virus disease would be conceivable. This would represent a particular risk for household spread. In addition, person-to-person transmission within health care facilities, as was seen during the 2014–2016 outbreaks in Africa and in the United States, would be not only possible but likely in the absence of rapid identification of the virus, immediate availability of isolation strategies, access to personal protective equipment, and procedures to protect health care workers.

Extensive preparation and specialized treatment facilities would be necessary to halt transmission and reduce the inevitable public fears that would be triggered by a release of the virus. In the absence of a vaccine, health care workers would have to receive thorough and repeated training in the use of protective equipment. Disposal of contaminated equipment, clothing, bedding, bodily waste, and other materials would require extensive capacity and planning.

During the West African epidemic, thousands of U.S. and European medical personnel went through training exercises to contend with the possibility of Ebola arriving at their facilities. This represented an enormous logistical burden, even in the absence of large numbers of cases.

HOW AN ATTACK MIGHT APPEAR

An unannounced attack with Ebola virus involving limited numbers of victims who had no known exposure would be difficult to detect at first. The illness caused by Ebola virus even in its most severe form shares many features with other infections that would be more likely outside of an outbreak setting and in a region not known to have EVD. Specifically, the progression of signs and symptoms described in Chapter 7, beginning with influenza-like features and culminating in a *cytokine storm*, would most likely be attributed to severe sepsis due to bacterial infection or malaria. The hemorrhagic features previously associated with EVD were not very common in the 2014–2016 West African cases and probably would not stand out sufficiently to suggest hemorrhagic fever specifically, perhaps until the number of cases reached a critical level. For these reasons, recognition that an attack with Ebola virus was underway probably would be delayed unless multiple cases presented to specific health care facilities in a short period of time without any other identifiable cause. Such a delay might be similar to that seen in the West African outbreak of 2014–2016. That outbreak was not recognized for several months after it began, and its true scope took even longer to be appreciated. It seems possible that EVD might be first suspected when illness was seen in secondary household cases or in health care workers or first responders involved in the care of the initial victims. The comparatively long incubation period, as long as 21 days, might further serve to obscure that an attack was underway and that Ebola virus was the cause.

POTENTIAL IMPACT OF AN ATTACK ON THE PUBLIC AND THE HEALTH CARE SYSTEM

As seen in the public reaction to the potential for international spread of EVD in 2014–2016, making it known that Ebola virus had been released into a community would be likely to result in fear out of proportion to the actual risk of contagion. This public reaction and the similar parallel reaction to be expected among health care workers might be more devastating to the health care infrastructure than the actual cases of EVD caused by the attack. Fears regarding novel routes of transmission, including by the airborne or respiratory droplet route or spread by inanimate objects (fomites) or casual contact, would likely grip large segments of the public, leading to calls for quarantine and other unwarranted attempts to limit the spread of infection. As in other attack scenarios, the *worried well* might flood emergency departments and physicians' offices asking to be evaluated for EVD. The absence of a rapid and easily available blood test and the need for potential quarantine and observation of patients who may have come into contact with a documented case could create enormous difficulties for the health care system. Local health departments would be called upon to provide diagnostic testing and to assist in developing plans for logistical problems such as disposal of biological waste. Such tasks and responsibilities would potentially distract from other important functions and require reallocation of resources to facilitate response to the attack. In addition, precautions required in handling clinical specimens from patients potentially infected with Ebola virus would tend to slow down laboratory operations in hospitals and public health facilities. Compounding this further would be the likely high rate of absenteeism of key health care workers and first responders, as has been projected in other biological attack scenarios and exercises.

All of these factors could have a profound impact on the local and national health care systems. The West African Ebola outbreak of 2014–2016 transfixed much of the world's population. Misinformation and exaggerated fears of contagion were prevalent for months in the developed and developing worlds, even under circumstances where the source of the outbreak was known and measures to contain it within the countries of West Africa were effective. A repeat of that reaction would magnify the impact of any intentional release of Ebola virus.

The treatment of EVD, as discussed in Chapter 3, requires an elaborate care environment for safe and effective treatment of the victim, as well as for protection of medical staff. For a widespread attack involving large numbers of victims to be conducted, many points of release of the agent, and, potentially, many suicide attackers, would be needed. However, the potential for spread to even a relatively small number of victims, as was seen in the United States in 2014, could lead to high levels of fear among the public and health care workers.

There is currently no established, available vaccine for EVD, as there is for smallpox and anthrax, and no proven effective antimicrobial therapy, as there is for plague and anthrax. This lack of a vaccine and of direct, effective therapy, other than supportive care, was responsible, in part, for the size of the 2014–2016 West African outbreaks. As described in Chapters 7 and 8, the treatment of a patient with EVD typically requires massive fluid resuscitation and treatment of electrolyte and clotting abnormalities. In some cases ventilator support and dialysis are needed. The fact that these measures must be provided by medical staff wearing extensive protective equipment and working in isolation facilities adds to the burden that even a relatively small number of cases of EVD could place on a health care system.

LIKELY RESPONSE TO A BIOLOGICAL ATTACK

As discussed in Chapter 9, very significant progress has been made toward an effective vaccine against EVD. If such a vaccine were widely available in the areas of Africa where Ebola virus is endemic, naturally occurring disease could be dramatically reduced. However, given the present circumstances in which EVD is not endemic anywhere outside of Africa, the stockpiling of a vaccine to be used in the event of a biological attack would not necessarily be regarded as feasible. In the event of an attack, ring vaccination, in which all close contacts of a patient would be vaccinated, could be an effective strategy. However, the logistics of accomplishing this might be formidable. In addition, even though the vaccines under development have not so far been associated with serious side effects, the acceptance of a vaccine might not be adequate to prevent transmission.

Whether or not a vaccine were available to counter an attack with Ebola virus, public health authorities would still play the key role in educating the public about the nature and symptoms of EVD and in minimizing the potential panic that an attack would be likely to cause. In addition, education of health care workers, with reinforcement of appropriate screening of patients for possible exposure to Ebola virus, would be a key strategy. Another extremely important public health mission would be educating and training health care workers about appropriate personal protective equipment and the proper ways to use it. Highly effective means of laboratory diagnosis of EVD would be essential and would require a coordinated effort between local, state, and national public health laboratories.

As can be seen from this brief overview, the impact of an attack using Ebola virus would be substantial, perhaps far beyond the actual numbers of victims. Fortunately, for the reasons indicated above, the logistics of effectively launching and sustaining such an attack would likely frustrate the effort. There has never been a recognized intentional release of Ebola virus. However, alterations in viral properties to facilitate more efficient means of transmission could one day render it a more feasible agent of attack.

REFERENCES

Cenciarelli O, Gabbarini V, Pietropaoli S et al. Viral bioterrorism: Learning the lesson of Ebola virus in West Africa 2013–2015. *Virus Res* 2015;210:318–326.

Cunha BA. Anthrax, tularemia, plague, Ebola or smallpox as agents of bioterrorism: Recognition in the emergency room. *Clin Microbiol Infect* 2002;8(8):489–503.

Gunaratne ND. The Ebola virus and the threat of bioterrorism. The Fletcher Forum of World Affairs. Winter 2015;39(1):63–76.

Kagan E. Update on Ebola virus and its potential as a bioterrorism agent. *Clin Pulm Med* 2005;12:76–83.

Masci JR, Bass E. *Bioterrorism: A Guide for Hospital Preparedness*. Boca Raton, FL: CRC Press, 361pp, 2005.

Passi D, Sharma S, Dutta SR, Dudeja P, Sharma V. Ebola virus disease (the killer virus): Another threat to humans and bioterrorism: Brief review and recent updates. *J Clin Diagn Res* 2015;9(6):LE01–LE08.

Frequently Asked Questions

11

The West African Ebola epidemic of 2014–2016 raised many questions about this previously rare infection. Much information was already known about the structure of the virus and the means by which it enters the body and causes human infection. In addition, information regarding animal reservoirs and vectors was known, at least to an extent. A surprising amount of work already had occurred on the development of both vaccines and antiviral medications. Prior to this outbreak, however, Ebola virus was thought to be one of the rarest human pathogens, and strategies to contain outbreaks as large as that seen in 2014–2016 were lacking. Concerns regarding potential global spread and, particularly, spread to the general population outside of Africa were essentially nonexistent. The virus was known as a highly lethal agent that posed risk to those living in somewhat remote areas of East Africa and, perhaps, to health care and laboratory workers responding to outbreaks in that region. What follows is a series of questions and answers both covering information known about Ebola virus species prior to 2014–2016, as well as discoveries and concerns that arose during that outbreak. In addition, possible future concerns are addressed. All of the material covered in this chapter in this format is presented elsewhere in this book in greater detail, and appropriate chapter references are provided. It is hoped that this format provides the reader with both easily understood information as well as a guide to the remainder of this book. The literature references to the information covered in this chapter can be found in the reference sections of the indicated chapters.

EBOLA VIRUS

Where Did the Ebola Virus Originate?

The first recognized human cases occurred in Zaire, now the Democratic Republic of Congo (DRC), in 1976 near the Ebola River. Whether the virus had caused human infections before that time is unknown. The similarity of the clinical manifestations to other hemorrhagic fever viruses and the limited capacity for identifying and reporting rare infections in the regions involved make it difficult to determine if this was a human pathogen earlier in the twentieth century. Of course, knowing whether Ebola had ever infected humans in earlier epochs is impossible. After 1976, East Africa saw 38 comparatively small outbreaks of Ebola virus disease (EVD), the largest of approximately 400 cases, primarily in the countries of Uganda, Gabon, Sudan, and DRC. Of the five species of the virus, two species—Ebola–Zaire (ZEBOV) and Ebola–Sudan (SEBOV)—caused all but one of these outbreaks. Almost all outbreaks of EVD prior to the 2014–2016 West Africa epidemics were brought under control in relatively short order. This sporadic, limited pattern of Ebola clusters probably reflected the fact that they began in more rural and less populated areas than the major cities involved in the West African epidemic. Since transmission typically requires direct contact to the body fluids of an infected individual, spread in rural areas involves fewer complicated, extensive chains of transmission than spread within an urban environment. The manner in which the small outbreaks prior to 2014 began was not always known, but exposure to infected animals, particularly bats or primates, was likely to have been the initiating event.

Chapter reference: 1

What Is the Animal Reservoir of Ebola Virus?

Similar to many human pathogens, Ebola virus exists in other species. They may form the reservoir that maintains the presence of the virus in a region and serve as the vectors of infection that transmit it to humans. A number of mammalian species may become infected with the Ebola virus. Extensive studies of wild animals in search of the reservoir of Ebola virus have identified only several species of fruit bats as carriers. These animals serve as vectors of transmission as well as when they are eaten or butchered. In addition, ingestion of vegetation contaminated with bat droppings may also represent a potential avenue of spread.

Primates, including monkeys, chimpanzees, and gorillas, may contract infection, and die-offs of such animals have been described in the vicinities of human outbreaks.

Chapter reference: 6

THE 2014–2016 WEST AFRICA EPIDEMIC

How Did the Epidemic Begin?

Retrospective studies have indicated that the West African outbreak began in late December 2013 when a two-year-old boy in Meliandou, a village in the Guéckédou region of Guinea, presented with fever and vomiting and died of what is thought to have been Ebola virus disease two days later. The source of infection has not been determined but is thought to have been wild animals, likely bats. In early January 2014, several family members of the boy, as well as staff at a hospital in Guéckédou, developed similar symptoms and died. Infection continued to spread to others who attended funeral rituals of victims. Other regions within Guinea saw similar illnesses as secondary and tertiary spread continued through multiple chains of transmission. The capital city, Conakry, saw its first case on February 1, 2014. The causative agent of these cases was determined to be Ebola virus in late March 2014.

Chapter reference: 1

What Was the Course of the Epidemic?

After its apparent beginning in Guinea in December 2013, it spread steadily to adjacent countries. It was introduced by travelers to the neighboring countries of Sierra Leone and Liberia. Spread then occurred through all the three countries by a variety of means. These included: infection of health care workers at treatment facilities and in the field, spread within families and social groups through burial practices, and other means. The virus also spread, via travelers, to three other countries, Nigeria, Mali, and Senegal, where limited outbreaks occurred. Due to the rapidity of spread, the involvement of urban areas (the capitals of all three countries) and a delayed governmental and international response, the magnitude of the epidemic was far greater than prior outbreaks. The epidemic peaked in late 2014 and subsequently came under almost complete control by January 2016, in large part due to dedicated resources from governmental and nongovernmental international relief efforts.

Chapter reference: 1 and Appendix

How Did the Epidemic become So Extensive?

The epidemic began gradually in the West African countries of Guinea, Sierra Leone, and Liberia but then accelerated quickly during the spring and summer of 2014 in all three countries. The rapidity of spread was the result of a number of factors. Among these were

- The early involvement of the densely populated capital cities of all the three countries. This fostered transmission within crowded populations and interfered with efforts to define and interrupt chains of transmission.
- Cultural practices, particularly those related to funerals, which could bring people in direct contact with the body or body fluids of deceased persons.
- Inadequate health care infrastructure, including doctor and nurse shortages and a lack of facilities to manage the intensive care needs of patients with EVD, as well as facilities to reduce the risk of contagion.
- In addition, inadequate systems of transportation.
- Fears of coming for care or transporting loved ones to treatment facilities because of concerns that death was more likely among those presenting for care.
- Obstacles to case findings and identification and interruption of chains of transmission.

Chapter reference: 1

What Were the Obstacles to Combating the Epidemic?

The countries in West Africa heavily impacted by the Ebola virus disease (EVD) outbreak of 2014–2016—Guinea, Sierra Leone, and Liberia—struggled to cope with the magnitude of this unprecedented event. A number of barriers to mounting an effective effort at containing the epidemic presented them-selves. Among these were

- A delay in recognition that the outbreak had begun because of inad-equate systems of diagnosis, surveillance, and reporting.
- Relatively rapid spread because many of the initial cases occurred in the densely populated capitals of the three countries.
- A severe shortage of physicians and other health care workers.
- A lack of appropriate containment facilities.
- An understaffed effort at contact tracing and community outreach.
- Funeral and burial practices that fostered transmission of Ebola virus infection through exposure to bodies and body fluids.
- Insufficient personal protective equipment and *logistical* difficulties of health care facilities to minimize the exposure of health care workers.
- Fears and skepticism regarding health care facilities as places to go for care.
- A delayed and slow response of governments outside of Africa in aid-ing the development of effective systems of treatment and containment.

In contrast to the three countries most heavily impacted, the neighboring coun-tries of Mali, Nigeria, and Senegal implemented effective measures to mini-mize the number of cases and saw a more rapid end of the outbreaks there. This was a result of more established and effective public health, laboratory, and clinical facilities, as well as the fact that the outbreak had already been recognized as Ebola in Guinea, Sierra Leone, and Liberia.

Chapter reference: 1 and Appendix

What Was the Global Response to EVD?

During the first few months of the Ebola epidemic in 2014, the world took little notice. For the first three months after the initial cases of an Ebola-like illness were seen in West Africa, there was uncertainty as to the cause. Because of its nonspecific symptoms and the fact that the countries involved, primarily Guinea, had not witnessed Ebola virus disease in the past, other more com-mon infections such as Lassa fever and cholera were thought to be the cause. When Ebola virus was documented to be the cause in March 2014, and the World Health Organization issued an alert, the seriousness of the outbreak became more apparent. But it was not until the late spring and early summer that the true scope of the epidemic became clear. As the world became edu-cated about the infection and Ebola became known as the highly dangerous infection that it is, the global reaction veered toward panic and concern for a pandemic that would not remain confined to Africa. Media coverage often ran ahead of the efforts of public health authorities to provide balanced educational information. Political concerns forced a scenario in which travel restrictions and calls for quarantine of returning travelers to the United States, Western Europe, and other regions in part distracted the public from the need for enhanced resources in West Africa to slow the epidemic and limit the humanitarian crisis that Ebola was causing in Guinea, Sierra Leone, and Liberia.

Scientific, medical, and public health authorities, both governmental and nongovernmental, from the developed countries faced challenges in tracking the dimensions of the epidemic while assisting the affected countries in developing effective strategies at contact tracing as well as diagnosis, containment, and treatment. The international aid efforts came in several forms.

Nongovernmental aid organizations, in particular Doctors Without Borders (MSF, for Médecins Sans Frontières), provided early and ongoing assistance and personnel to the impacted countries. Governmental agencies, including the U.S. Centers for Disease Control and Prevention (CDC) as well as comparable entities from the United Kingdom, Western Europe, and other developed countries became increasingly involved in the efforts on the ground during the summer and fall of 2014. The U.S. Military developed and maintained a presence and constructed a number of treatment units. While providing essential epidemiologic and scientific support, the World Health Organization (WHO) was criticized by some for what was considered a delayed response to the crisis.

As the response on the ground to the emerging epidemic accelerated and broadened, efforts to develop a vaccine and to investigate new antiviral therapies intensified.

Overall, despite the delay in the recognition of Ebola and the deployment of sufficient resources to combat its spread, the epidemic had begun to be brought under control through the efforts of many organizations, both local and international, private and governmental, by the end of 2014.

Chapter reference: 4

EBOLA VIRUS DISEASE: PATHOGENESIS AND TRANSMISSION

How Do Humans become Infected?

The most important route of transmission to humans is by direct contact with the body fluids of infected individuals. In the West African outbreak of 2014–2016, this was the means by which many individuals, particularly health care workers and those attending burial rituals of deceased victims contracted the infection. Speculation that infection could be transmitted by the respiratory route was not borne out by the patterns of spread. This speculation, however, did contribute to anxiety and fears that Ebola infection could be acquired simply be being in close proximity to an infected person without physical contact, thus raising concerns regarding the safety of health care facilities and a host of other environments, including airports, planes, passenger vehicles, and other confined spaces. Sexual transmission, however, was documented for the first time during this outbreak. The frequency of transmission by this route is unknown. Identifying instances in which infection was contracted sexually during the 2014–2016 period is difficult because the far more likely route of spread by direct contact with other body fluids may have obscured occasional sexual transmission. The potential importance of sexual transmission is great because it may represent a means by which infection could continue at a lower but continuing rate. As discussed below, this concern has led to recommendations to avoid sexual contact following EVD for a period of several months.

Chapter reference: 6

How Is Infection Transmitted from Person to Person?

The two routes of person-to-person spread are by direct contact of the skin or by mucous membranes with body fluid from infected individuals or corpses. There are multiple lines of evidence that this is by far the major route of transmission. Speculation regarding the potential for airborne spread was proven unwarranted by the documented chains of transmission. Airborne spread, if possible, would have been expected to result in a much larger outbreak, affecting individuals not in physical contact with victims. There was no evidence that this occurred, despite the opportunities for such spread in the crowded urban areas of Guinea, Sierra Leone, and Liberia were the most heavily impacted. Prior work had demonstrated that transmission through inanimate objects such as furniture and utensils was rare or nonexistent. As in the case of airborne transmission, this route would have likely been revealed through contact tracing. The efficiency of spread through direct contact with the body fluids of severely ill patients was clearly demonstrated in two instances of transmission to nurses working at a Dallas hospital and caring for a man who was severely symptomatic and subsequently died of EVD. These nurses were using personal protective equipment (PPE) and the exact nature of the exposure to skin was never characterized in published literature. These cases led to additional guidance and more structured training in the use of PPE. Evidence for ready transmission by direct contact with bodies and body fluids of infected individuals was present throughout the epidemic in West Africa as health care workers and individuals involved in the treatment and care of corpses of victims were two prime risk groups for infection.

Sexual transmission was documented for the first time during the 2014–2016 epidemics, when a man who had apparently recovered from acute infection transmitted infection through sexual intercourse. The possibility that sexual transmission had played a role during the outbreak, and the risk that it could cause additional cases after the end of the epidemic, are currently being examined.

Chapter reference: 3 and 6

What Is the Evidence That Infection Can Be Spread by Sexual Contact?

In studies prior to the 2014–2016 West Africa epidemic, evidence of Ebola virus had been found in the genital fluids of infected individuals, particularly the semen of men, after viremia had resolved. But sexual transmission had not been documented until the West African epidemic, when a Liberian woman appeared to have contracted Ebola infection by sexual contact with a man who had recovered more than five months earlier. The role and importance of this route of transmission is not yet clear; however, persistence of virus in this manner raises two important concerns:

- Sexual transmission may represent a means for Ebola virus to remain endemic in a region after an outbreak has been brought under control.
- Persistence of virus in genital fluids may lead to late recrudescence of symptoms after initial recovery.

On the basis of these concerns and the case described above, the World Health Organization issued a guidance regarding sexual transmission, which recommends that sexual intercourse should be avoided until two semen specimens test negative for the virus or six months has elapsed since the male partner has recovered from EVD.

The possibility of sexual transmission remains a difficult issue. Strategies to establish that all infected men have cleared the virus from their semen before having sexual contact have not yet been developed.

Chapter reference: 6

How Long Can an Infected Individual Remain Contagious?

In general, the period during which a patient with EVD may transmit infection correlates with the period of viremia. This typically lasts for one to two days after the infection first becomes symptomatic, but may last longer. Transmission by nonsexual contact is not thought to be possible until an infected individual becomes symptomatic. Chances of transmission decrease rapidly with clinical recovery. However, longer-term carriage in the semen was documented for the first time during and after the 2014–2016 West Africa epidemics, and transmission by sexual contact appears to have occurred from a man to a woman more than five months after he recovered from acute infection. The frequency of this form of chronic carriage of the virus and the risk of transmission has not yet been fully defined.

Chapter reference: 6

Can the Infection Be Spread through the Air?

During the 2014–2016 West Africa epidemic, concerns were expressed that Ebola virus might mutate to forms that could spread through the air or by respiratory droplet. There is no evidence that Ebola virus can be transmitted by the airborne route or by inhalation of respiratory droplets. Moreover, the pattern of transmission of disease and the relatively limited number of cases in crowded environments provide evidence against spread by the respiratory route. In addition, there is no known instance in which any other pathogenic virus has mutated to become an airborne pathogen. A number of issues would seem to make likelihood of mutation to an airborne form extremely remote. For example, multiple mutations probably would be required to cause the virus to reach high levels in respiratory secretions from which it could be readily released. Additional alterations would be needed to permit the virus to remain viable as particles in the air before reaching the next host in an infectious form in sufficient quantity to cause disease. It should be noted that mutations are both random and not universally beneficial, so that any mutations that might make airborne transmission more likely might also be deleterious to the virus. Despite these points, any indication of novel routes of transmission of Ebola virus must be elucidated thoroughly in light of the highly lethal infection that it causes.

Chapter reference: 6

How Has It Been Determined That Infection Cannot Be Spread through Fomites (Inanimate Objects)?

There is little evidence that transmission can occur by contact with inanimate objects unless they are contaminated with the body fluid from infected individuals. In a study conducted before the 2014–2016 outbreak, Bausch and colleagues found that virus was rarely (2/33) detected on inanimate objects and that recommended infection control procedures would probably prevent transmission from these objects to health care workers.

Chapter reference: 3

Can Chronic Infection Occur?

Prior to the West African epidemic of 2014–2016, no recognized cases of chronic infection with Ebola virus had been seen, and no potential reservoir for persistent viral replication within the body was known. However, persistent infection after apparent recovery has been documented in several sites in the bodies of some patients. These include the vitreous humor, cerebrospinal fluid, and the sexual glands of men. The maximum duration and potential significance of persistent infection and the likelihood of systemic relapse is not known at this time. Further, the potential role of transmission by sexual contact within the West African outbreak has not been defined.

Chapter reference: 6

Can Ebola Virus Infection Be Asymptomatic?

Asymptomatic Ebola virus infection appears to occur. Recent analysis of outbreaks prior to the 2014–2016 West Africa epidemics suggests that the rate of asymptomatic infection is substantial. Although these patients cannot transmit infection, it is possible that asymptomatic infection confers protection against future symptomatic infection. The presence of subclinical infection may explain the differential attack rates seen within households. It may also provide an explanation for the observation that some individuals, particularly young children, seem relatively resistant to infection.

Chapter reference: 6

Can Ebola Virus Persist in the Body after Clinical Recovery?

Ebola virus had not been known to persist in the body, other than in genital fluids, prior to the 2014–2016 epidemics and its aftermath. Evidence of persistence of infection has appeared in a small number of patients who have recovered from acute EVD, and evidence of infection has been detected in the cerebrospinal fluid (CSF) and the vitreous of the eye months after the resolution of viremia. Because of the few cases in which these forms of persistence of infection have been documented, the significance and meaning of this phenomenon are not fully understood. Of interest, two of the patients, both health care workers from Europe who had contracted EVD while working with victims in West Africa, had received immunotherapy for their acute infections, raising the concern that this form of therapy may promote viral persistence and, potentially, clinical recrudescence with recurrent viremia arising from these *protected sites*.

Chapter reference: 6

What Progress Has There Been toward a Vaccine?

As the magnitude of the 2014–2016 epidemics became evident, increasing attention fell on the efforts to create an Ebola vaccine. The earliest efforts at creating a vaccine began shortly after human cases were seen for the first time in 1976. For years prior to the West African outbreak, a number of vaccines had been under investigation. With its high mortality, Ebola virus was regarded both as a regional and potentially global hazard. In addition, concerns had been raised that the virus, along with hemorrhagic fever agents, could be used as a biological weapon and as a means of conducting bioterrorism. The sudden outbreak in Guinea, Sierra Leone, and Liberia of thousands of cases of an infection that had been considered dangerous but rare lent increased relevance to efforts at containment. With the fear that a global pandemic could be on the horizon, the sense of urgency to develop an effective vaccine grew dramatically.

A number of vaccines were already under evaluation in animal models, including monkeys and rodents, when the outbreak began. The U.S. Food and Drug Administration's so-called *animal rule* allows vaccines to be licensed on the basis of animal data if human testing is not feasible, especially in cases of highly lethal diseases such as Ebola. This rule was applied by the FDA for the first time in 2015, when it licensed the anthrax vaccine BioThrax.

Inactivated virus and viral subunit vaccines were of inconsistent effectiveness, and subsequent work employed viral vectors.

Remarkably, by mid-2016, a vaccine using a recombinant cattle virus, vesicular stomatitis virus (rVSV–ZEBOV), which produces an Ebola protein, was demonstrated to confer 100% protection in humans. However, as discussed in Chapter 9, the study of this vaccine was limited because at the time of the trial there were relatively few cases occurring.

Complex ethical and logistical issues are raised by the possibility of an effective Ebola virus. These center on how the vaccine can be most effectively used. Since the vaccine is new, unanticipated safety issues and efficacy issues may emerge that will require systematic follow-up. This level of ongoing monitoring may be logistically impossible in some of the countries where Ebola is most likely to reemerge. However, simply providing vaccine to health care volunteers or public health or military personnel from the developed countries can certainly be viewed as discriminatory against populations in Africa at ongoing risk.

Chapter reference: 9

What Is *Ring Vaccination*?

Ring vaccination is a containment strategy in which concentric rings of contacts of an infected person are traced, identified, and vaccinated. The initial ring typically may include family, neighbors, and coworkers. Then contacts of the inner ring are identified and vaccinated, and so on. It is a form of containment rather than primary prevention. This strategy was used successfully to eradicate smallpox in the 1970s, considered one of the most important successes of the twentieth century public health measures. It was also used in the trial of the rVSV–EBOV Ebola vaccine trial whose results were published in late 2016.

Chapter reference: 9

EBOLA VIRUS DISEASE: CLINICAL FEATURES AND MANAGEMENT

What Is the Incubation Period of Ebola Virus Disease?

The incubation period is typically between 1 and 21 days, with the average eight days. Patients experiencing short incubation periods often have more severe infection and higher rates of mortality. The variability in the incubation period is likely to reflect the amount of virus (inoculum) in the body fluid to which exposure occurred. Patients with longer incubation periods may have milder disease. The relatively long asymptomatic period, however, can provide an opportunity for infected patients to travel and represent the first case in a new cluster remote from the area where they acquired the infection.

Chapter reference: 7

What Are the Initial Signs and Symptoms of EVD?

Infection begins with nonspecific symptoms. After an incubation period lasting from 1 to 21 days (average eight days), early symptoms begin with fever, which occurs in approximately 90% of the cases, followed by muscle aches, headaches, and gastrointestinal complaints. The features of infection are sufficiently vague and nonspecific to cause confusion with a host of other infections, particularly malaria, most of which are more common than Ebola (see below).

Chapter reference: 7

What Is the Differential Diagnosis of Ebola Virus Disease?

The differential diagnosis of these nonspecific features is broad and includes many infections that are more likely than EVD in Africa. Among these are malaria, measles, dengue, leptospirosis, bacterial sepsis, typhoid, and influenza.

Chapter reference: 7

How Does the Infection Progress?

In the typical case of EVD, after three to five days of nonspecific symptoms, gastrointestinal manifestations, including diarrhea, nausea, and vomiting, begin to dominate. With the onset of gastrointestinal losses, viral shedding reaches a peak, and the risk of contagion to others caring for the patient is great. Due to the resultant volume depletion and severe sepsis physiology, hypotension and renal failure can then occur. Hemorrhagic complications were seen less frequently in the 2014–2016 outbreaks than in previous clusters, but a bleeding diathesis manifested by thrombocytopenia can be significant components of the illness in some individuals. Bleeding from the gastrointestinal tract, the lungs, the nose, the gums, and the conjunctivae, as well as bleeding from injection sites, may be seen. Nonspecific skin rashes, similar to those seen in measles, parvovirus B19, enteroviral infections, and rubella, have frequently been reported in association with Ebola infection in prior outbreaks, although this manifestation was seen in only approximately 5% of cases in the 2014–2016 West Africa epidemic.

As illness progresses, hypotension may supervene and may be accompanied by pulmonary edema as a manifestation of capillary leak. The accompanying physical findings include evidence of severe dehydration with dry mucous membranes and orthostasis. Progressive pulmonary involvement resulting from pulmonary edema may be evident on examination of the lungs.

In fatal cases, death typically occurs during the second week of illness after a period of profound fluid loss and end-organ damage, which might include renal and central nervous system involvement. Hemorrhagic complications may also contribute to death. Fulminant hepatic necrosis may become clinically apparent and is a frequent postmortem finding.

Chapter reference: 7

What Are the Laboratory Features of Ebola Virus Infection?

The laboratory findings of Ebola infection typically include the following:

- Leukopenia is commonly seen with prominent depletion of lymphocytes.
- Atypical lymphocytes may become apparent and the infection progresses.
- Thrombocytopenia is almost always present and, in fatal cases, typically persists until death.
- Prolongation of the prothrombin and partial thromboplastin times may be seen. D-dimer may be elevated as a reflection of disseminated intravascular coagulation.
- Nonspecific elevations of transaminases may occur, although jaundice is uncommon.
- Elevations of BUN and creatinine are typical. Acute renal failure may occur as a result of volume depletion and refractory hypotension.
- Hypokalemia reflecting gastrointestinal losses of potassium is common.
- Several laboratory markers of poor prognosis have been suggested. These include: elevated levels of ferritin and thrombomodulin;

four-fold or greater elevation of D-dimer; elevated levels of aspartate aminotransferase (AST); elevations of creatinine, BUN, and amylase, and decreased levels of serum albumin and calcium.

Age over 30 and extreme levels of fatigue, as well as central nervous system depression, may also be poor prognostic factors.

Virus can be detected in the blood at the onset of symptoms, often remaining low for several days and then rising, before ultimately falling with clinical recovery. In fatal cases, viral levels remain high and often reach their highest levels at the time of death and afterward in body fluids. Virus has been detected in saliva, stool, breast milk, tears, and seminal fluid, as well as in blood.

Chapter reference: 7

What Are the Principles of Management?

Correction of fluid losses due to gastrointestinal losses and the resultant hemodynamic instability is the mainstay of therapy. Secondary pulmonary involvement may necessitate respiratory support and mechanical ventilation.

Renal replacement therapy, including hemodialysis, may become necessary because of the renal insult resulting from hypotension. The role of investigational antiviral therapy and immunotherapy has not been fully established. Since most of the patients who received these measures in the 2014–2016 West Africa outbreaks were treated in the United States or Western Europe, it is not clear to what extent the improved survival of those patients reflected aggressive volume replacement and critical care rather than specific effects of antiviral or serotherapy.

Chapter reference: 8

What Is the Role of Antiviral Therapy?

The mainstay of management of EVD is adequate replacement of fluids and electrolytes, hemodynamic stabilization, antidiarrheal measures and respiratory support and renal replacement therapy if needed. The 2014–2016 West Africa epidemic stimulated interest and research into more specific antiviral therapy. Several strategic pathways of such therapy for Ebola infection have been explored. Although some hold promise, only a handful of patients have received these agents, none of which has yet been accepted as effective and gained approval by the United States Food and Drug Administration. The patients treated with antiviral therapy have been in settings in the United States and in Europe where intensive care could also be provided. Thus the role of antiviral therapy in the patients who received it cannot yet be distinguished from the benefits of other components of critical care.

Several drugs interfere with viral replication. These include the following:

- *Ribavirin*: This agent has long been used to treat a variety of viral infections, including hepatitis C and respiratory syncytial virus pulmonary infection. Ribavirin has been used successfully in the treatment of hemorrhagic fever viruses, including Lassa fever and Crimean–Congo hemorrhagic fever virus, but has been ineffective in Ebola virus infections.
- *Brincidofovir*: Brincidofovir is a prodrug that is metabolized to the antiviral agent cidofovir after cell entry. Cidofovir is active against cytomegalovirus and other DNA herpesviruses and has also been used in the treatment of the JC virus infection progressive multifocal leukoencephalopathy (PML). Despite the fact that Ebola virus is an RNA and not a DNA virus, brincidofovir has been demonstrated to inhibit its replication by an unknown mechanism. Several patients with Ebola were treated in the United States with this drug during the 2014–2016 West Africa epidemics. Although patients treated in the United States and Europe for Ebola infection had a higher rate of recovery than those, in general, treated in Africa, the correlation with outcome in the few patients who received brincidofovir has, so far, not been clear.

ZMapp, a blend of three chimeric monoclonal neutralizing antibodies directed at Ebola virus, received, perhaps, the most attention during the 2014–2016 outbreaks. It had previously been shown to be effective in a monkey model of Ebola infection. It was administered to a small number of patients during the West African outbreak. As was the case with brincidofovir, the effect of ZMapp on outcome was impossible to judge because of the limited number of patients who received the agent and the fact that they were treated in modern hospitals with critical care capability.

In light of the limited proven therapeutic options, research continues into strategies to reduce or eliminate viremia, including novel methods such as plasmapheresis.

Chapter reference: 8

What Is the Role of Immune Serum?

As in the case of antiviral therapy, serotherapy has an uncertain value. During the West African epidemic, serotherapy took the form of infusion of serum containing antibodies against Ebola virus from patients who had recovered. Because of the simultaneous provision of modern intensive care techniques in these cases, a clear distinct benefit from this form of therapy cannot yet be discerned. This type of therapy is also reviewed in Chapter 8 of this book.

Chapter reference: 8

What Is the Appropriate Approach to Fluid Management?

Resuscitation with oral and intravenous hydration is critically important in the management of EVD. Because of gastrointestinal losses due to diarrhea and vomiting, coupled with the *cytokine storm* and its attendant hemodynamic changes, hypovolemic shock is, perhaps, the main contributor to death in patients with advanced infection. As in the case of gram-negative sepsis, vascular tone and peripheral vascular resistance is dramatically reduced, leading to a substantial volume deficit. Failure to maintain intravascular volume through the administration of intravenous isotonic fluids risks acute renal insufficiency and other manifestations of end-organ ischemia. Of course, fluid administration may have negative consequences as well because capillary leak may result in early pulmonary edema and the need for respiratory support.

Chapter reference: 8

How Should the Response to Treatment Be Followed and Interpreted?

Hemodynamic stabilization through fluid resuscitation and vasopressor therapy is the mainstay of therapy. For this reason, blood pressure, heart rate, renal function, urine output, and neurological status, all direct or indirect indicators of a response to this therapy are the key parameters. Blood coagulation parameters, particularly platelet counts, also give an indication of progression of infection or response to therapy. Where possible, viral load in the blood provides a direct indication of recovery or worsening and can be used to monitor therapy with or without the use of antiviral medications. Indirect measures of inflammation, such as erythrocyte sedimentation rate (ESR), C reactive protein (CRP) levels, and serum ferritin levels, may also offer insight into the clinical status and may be predictive of subsequent improvement or worsening.

Chapter reference: 8

What Are the Features and Consequences of Ebola Infection during Pregnancy?

Ebola virus infection during pregnancy is particularly devastating, with a reported mortality of approximately 90%. Transplacental fetal infection appears to be common, and Ebola virus has been detected in placental tissue and amniotic fluid. Pregnancy outcomes are similarly discouraging, with a high rate of spontaneous abortion and universal or nearly universal death of infants born to infected women within the first few days of life. Ebola virus has been found in breast milk. Clinical manifestations among pregnant women are similar to those seen in other patients with diarrhea and vomiting and the resultant fluid losses dominating, often with progression to vascular collapse in fatal cases. As in nonpregnant patients, hemorrhagic complications are seen occasionally.

Chapter reference: 7

What Are the Consequences of Ebola Infection in Children?

The incidence of infection among young children in the 2014–2016 epidemics was relatively low, as it had been in prior outbreaks. For this reason, relatively little information has been published on the manifestations of the infection in children and the outcome of therapy. This may indicate that children were largely spared from direct contact with sick and dying adults, a common mode of transmission of Ebola infection to adults. The limited data comparing response to infection in children and adults suggests somewhat different, and perhaps more vigorous, immune response in children. However, the clinical manifestations in children are similar to those in adults. Symptoms typically begin with fever, headache, myalgia, and abdominal pain, with progression to diarrhea and vomiting, and in the minority of cases, hemorrhagic manifestations. The high mortality rate is also similar to that seen in adults. Neonates born to women with EVD have seldom survived longer than several weeks and the rate of fetal and neonatal infection appears to be nearly 100%. Since virus has been detected in breast milk, postnatal infection may be possible.

Chapter reference: 7

INFECTION CONTROL IN HEALTH CARE FACILITIES

What Is the Risk of Contagion to Health Care Workers?

As reviewed in Chapter 1, both clinical and nonclinical health workers were at extraordinarily high risk of contagion in the 2014–2016 West Africa outbreaks. This risk was between 20 and 30 times greater than the risk in the general population, with comparably high mortality of approximately 65%. This heavy toll was particularly devastating to the fragile health care systems in Guinea, Sierra Leone, and Liberia. The risk was borne disproportionately by nurses and nurse's aides, particularly in the early stages of the epidemic.

Any health care worker likely to come into skin or mucous membrane contact with body fluids from an infected patient is at risk of acquiring EVD. The potential categories of health care workers at risk in this way include any clinician or technician providing direct care, as well as laboratory personnel processing body fluid or tissue specimens. In addition, maintenance as well as transportation and mortuary workers and security staff may be at risk for this type of exposure.

Outside Africa, the only instances of transmission of EVD occurred among health care workers. Two nurses in a Dallas, Texas hospital contracted infection when caring for a patient from Liberia. This occurred despite the fact that the patient was confirmed to have Ebola infection and personal protective equipment was utilized. In addition, a nurse's aide in a Madrid, Spain hospital contracted EVD after caring for a missionary physician who had been evacuated from Sierra Leone and was known to have Ebola. In these cases, both of the patients infected in West Africa died, whereas all three health care workers survived.

Chapter reference: 1, 2, 3, and Appendix

How Can the Risk of Transmission to Health Care Workers Be Minimized?

Several measures are essential to reducing this risk. The first is effective identification and triage of individuals presenting for care. A history of presence in an area of active transmission should be sought in all patients during a regional or global outbreak of EVD. Health care workers having initial contact with a patient whose travel history and symptoms suggest the possibility of EVD should avoid physical contact with the patient and escort them to an isolation room, if available, for further evaluation. If there is no ability to provide this, contact between staff, visitors, and other patients should be prevented by establishing a perimeter of at least three feet around the patient. Patients with *wet* symptoms, for example, vomiting, diarrhea, or hemorrhage, may require a larger perimeter.

All patients, with or without symptoms, should be contacted only by staff wearing appropriate personal protective equipment (PPE), as described below.

The same precautions of isolation and PPE use should be maintained until patients are either proven not to be infected with Ebola or until they have recovered.

Once a patient is suspected to have EVD, health care workers need to wear personal protective equipment that fully covers all skin and mucous membrane and need to follow rigorous procedures in donning and doffing this equipment. The patient's wastes need to be disposed of in a way that prevents contamination of staff or the environment.

Chapter reference: 3 and Appendix

What Are the Principles of Personal Protective Equipment?

Ebola virus infection is transmitted when skin of mucous membranes come into direct contact with the body fluids of an infected individual. It has been assumed that contact between such fluids and intact epidermis is unlikely to result in transmission and that microscopic breaks in the skin are needed. For this reason, it was thought that the routine measures to prevent contact exposure to health care workers caring for a patient with EVD would be sufficient to prevent transmission. However, viral concentration in body fluids, including surface fluids such as sweat, may reach extremely high levels in the later stages of severe EVD and after death. As a result, it is assumed that relatively limited contact may be sufficient for transmission. Two nurses caring for a patient dying of EVD in a Dallas, Texas hospital during the 2014–2016 epidemic contracted infection despite using personal protective equipment. The principles of personal protection that evolved during the outbreak are based on completely covering all exposed skin and mucous membranes with impermeable personal protective equipment. The complexity of proper donning of such equipment has led to the recommendation that individual health care workers must receive specific and regular training in the procedure, and, ideally perform donning under the supervision of a trained observer. Since body fluids from the patient may contaminate the surface of PPE during care activities, removing the equipment (i.e., doffing) may carry an even greater risk of transmission of infection. For this reason, doffing, too, should be carried out only by personnel who have received specific training if possible. Like the donning procedure, doffing should be done with the assistance of a trained observer. The equipment used must be disposed of as contaminated, biohazardous waste.

Chapter reference: 3 and Appendix

PLANNING/LESSONS LEARNED

What Are the Lessons of the 2014–2016 Ebola Epidemic?

A number of lessons were learned both within the countries most severely impacted by Ebola virus disease (EVD) and in the rest of the world. Routes of transmission were more clearly documented than previously and the barriers to contact tracing and, ultimately, to containment were demonstrated dramatically. In the developed world, the needs of resource mobilization to contain outbreaks in resource-poor settings were illustrated, as they had been in previous epidemics of novel pathogens. The mobilization of strategies of vaccine development was also accelerated. As had been witnessed in prior frightening infectious disease threats, the impact of public concern and the needs of clear educational efforts were great.

The global realities of health care disparities were on display throughout the 2014–2016 West Africa Ebola epidemics. It remains to be seen whether the lessons of this epidemic will be used to limit the scope of similar outbreaks in the future. However, a number of specific lessons are of particular importance. Among these are

- When the arrival of EVD can be recognized, early and effective contact tracing can be implemented quickly, and large-scale outbreaks may be prevented. This was demonstrated in Nigeria, Mali, and Senegal, all countries neighboring the epidemic countries in West Africa, where containment measures prevented the most ongoing transmission.

- PPE use needs to be extremely meticulous to protect health care workers.
- Modern supportive care, including in some cases artificial ventilation and dialysis, as was used in patients evacuated to the United States and Europe, appears to be effective in saving lives.
- Timely distribution of accurate and culturally appropriate information to the public and health care workers is important in promoting behaviors that can reduce the risk of transmission and increase compliance with recommended public health measures.

Chapter reference: 1, 3, 4, and Appendix

How Should Health Care Facilities in Developed Countries Prepare?

Because of the high risk of contagion to health care workers seen in the West Africa EVD outbreak of 2014–2016, extensive measures have been developed in the United States for facilities anticipating that they might provide care to patients with EVD. The main thrust of these measures is to prevent health care workers from becoming infected by means of direct contact with the patient's body fluids. This contact may occur in direct patient care activities, relatively invasive procedures (e.g., endotracheal intubation), nonmedical physical contact with the patient, as well as primary contact with body fluids, as it might pose a risk to the laboratory, housekeeping, or maintenance personnel. Postmortem contact by mortuary personnel may also pose a significant risk of transmission.

For these reasons, adequate preparation should take into account the role that the facility could potentially play in the care of the patients as well as the types of patients likely to be encountered.

Initial triage and isolation of a patient thought to be potentially infected with Ebola virus must include the capability of providing a single isolation room in which the patient can be evaluated. The personnel trained in the proper use of appropriate personal protective equipment must be available, as must adequate PPE in appropriate sizes. A means must be developed for proper disposal of body fluids, and strategies must be in place for conducting routine blood and urine tests without exposing laboratory and transport workers to unnecessary risk. Arrangements must be in place for testing the blood for Ebola virus infection. This would typically include cooperative strategies with local and state public health laboratories for transport of specimens. Even for this basic level of care, preexisting plans are required and training of key personnel at regular intervals is necessary. The specific logistics of moving patients to isolation rooms (e.g., the appropriate routes to be used within the facility) must be included in the plan. If the patient is to be transferred to a different facility for biocontainment, the means of arranging and carrying out this transfer should be explicitly addressed. The potential disposal of bodies must also involve communication in advance with local medical examiner personnel.

During the 2014–2016 West Africa epidemic, hospitals throughout the world expended remarkable amounts of time and resources to prepare for patients with Ebola virus disease. Going forward, there will be a need for more universal, all-hazard approaches to outbreaks of novel pathogens both to prepare for the safe and effective care of patients suffering from these infections and also to preserve all other hospital functions.

Chapter reference: 3 and Appendix

What Should the U.S. Public Be Concerned about in Outbreaks of EVD?

Naturally occurring outbreaks of EVD probably will again begin in Africa. As was learned in the 2014–2016 epidemics, EVD poses little risk of becoming a global pandemic. There are multiple reasons for this, among these are

- Transmission of Ebola infection requires close contact with body fluids. This fact limits spread within communities to, for example, household and sexual contacts, health care workers, funeral workers, and so on. The absence of airborne, insectborne, foodborne, or waterborne transmission, as well as the unlikely transmission by fomites drastically limits the risk of spread within communities in the United States, even if infected travelers enter the country.
- Transmission can occur only after the onset of symptoms. Since intercontinental travel with symptomatic EVD would be unlikely, this fact limits the likelihood of substantial numbers of infected travelers entering the United States.
- Health care facilities in the United States are well equipped to limit contagion within a hospital. Hospitals typically have isolation rooms, personal protective equipment, and infection control procedures designed to provide the safe evaluation of patients with contagious disease, including many that are far more contagious than EVD.
- Public health facilities, although variable by state and municipality, are equipped to provide assistance in the processing of diagnostic

specimens. This capability is likely to result in more rapid diagnosis and tracing of contacts in the event that a traveler with EVD arrives.

- Improved strategies for containing outbreaks of EVD and similar pathogens will, presumably, greatly reduce the risk that a regional epidemic will spread far beyond the original affected zone.

Despite these reassuring facts, as was learned in the 2014–2016 EVD epidemics and in outbreaks of other special pathogens, the actual low risk of contagion may be lost in the high level of public anxiety. For this reason, it remains critically important for public health authorities to be able to disseminate accurate information to the public and for media outlets to serve as partners in this process. The internet offers never-before-seen opportunities for mass communication of accurate information. This capacity, if used effectively to make accurate information widely available to the general public, may reduce the sensationalism and excessive politicization witnessed during the EVD outbreak of 2014–2016. Unfortunately, false or misleading information may also be distributed in this manner.

Chapter reference: 3 and 6

How Should the Public Be Kept Informed?

Studies have shown that the source of information that the public relies on most during these events are their own health care providers.

Health care professionals and public health authorities, however, often become aware of these events on short notice. What was seen during the 2014–2016 Ebola epidemics in West Africa was a largely exaggerated and unnecessary wave of fear across the general population of many countries that had little reason to expect to see cases of the infection and even less reason to anticipate transmission of Ebola within their country. Nonetheless, a combination of inflammatory and inaccurate media coverage and often ineffective communication to the public by medical and public health authorities resulted in levels of concern that bore little relationship to the actual risk.

The challenge of preventing such a drastic overreaction in circumstances like those during the West African outbreak is great. A novel, exotic pathogen that causes a disease with a high mortality rate for which there is no specific treatment and no vaccine can be perceived as a risk for the entire world's population unless certain facts are communicated clearly and repeatedly. In the case of Ebola, these facts were the following:

- The likelihood of infection is essentially nonexistent outside the area where active transmission is occurring.

- Transmission requires close, essentially intimate, contact with an infected person.
- An infected person is capable of transmitting the infection only when he or she has symptoms.

Since individuals rely on their own health care providers for information regarding public health risks, the fears of many in the health care professions regarding their own possible risk may have interfered with the provision of accurate, nonemotional information emphasizing these points.

Chapter reference: 4 and 6

Tabletop Exercises for Preparedness

12

INTRODUCTION

A significant challenge for health care systems in preparing to provide care for a patient with Ebola virus disease (EVD) is training staff members to respond in ways that may be unorthodox. Individual duties may change, and interdisciplinary teamwork may take on new dimensions. Because there is a risk of transmission of Ebola virus to health care workers as they perform their duties, and because EVD is a serious and frequently fatal infection, high levels of anxiety are likely to exist among potential caregivers. For these reasons, careful planning for the various types of scenarios that EVD may present in a health care facility should include education of staff at all levels and opportunities for those likely to be involved to ask questions and express concerns. In addition, organizational planning should take into account the novel challenges that staff fears and public and media concern and curiosity may pose.

When the arrival of a patient with EVD becomes a possibility, these issues need to be explored: the need for rapid identification of patients with possible or likely EVD; the steps needed to prevent transmission of infection to staff and other patients; the specific roles of each individual involved, and the organizational strategies required to effectively deal with press inquiries, the worried well, employee absenteeism, shortage of protective equipment, and perhaps, uncertain guidance from public health officials.

Tabletop exercises have been used by health care organizations to prepare for a host of potential incidents. These include weather emergences, power failures, mass casualties, terrorist attacks, active shooters, and other incidents. In recent years, special pathogens, including not only Ebola but pandemic influenza, novel *coronavirus*, and the pathogen that causes severe acute respiratory syndrome (SARS), among others, have necessitated planning using these types of exercises.

USING TABLETOP EXERCISES

The tabletop exercises presented here are designed for health care facilities in the United States or other countries with ample medical resources. They address the evaluation of various types of patients: (1) a nonhealth care-related traveler, (2) a symptomatic student volunteer returning from a country affected by EVD, (3) an asymptomatic health care worker returning from work in an Ebola treatment unit in an endemic country, (4) a possibly exposed individual presenting to an ambulatory care site, and (5) a pregnant woman with possible EVD. Each consists of a multipart scenario describing possible situations in which the evaluation, triaging, and initial steps on management of the patient are explored. The intended core audience for each exercise is defined. These exercises can form the basis for a presentation and discussion with the suggested participants. The aim is to help to clarify the strategies to be used in these situations. Surveys of the knowledge of primary care providers about these issues have tended to suggest important gaps in understanding, in the United States (Ganguli et al. 2015; Highsmith et al. 2015) and Europe (Valerio et al. 2015).

The structure of exercises such as these permits opportunities for discussing the issues identified in great depth, when necessary. In addition, the format allows for the exploration of various specific factors that can be introduced during the exercise, so-called injects. Injects may be designed for the specific audience and participants of the exercise. For example, an exercise focusing on the emergency department (ED) triage and protective measures relevant to a returning traveler with fever could also address an issue such as communication with the media. This aspect of the scenario can be amplified by adding an inject, such as this piece of information: "As the patient is being evaluated, a call comes from a reporter for a local TV station who has learned that a suspected case of Ebola infection is under evaluation and who would like to speak to the physician caring for the patient." Such an inject should raise issues regarding patient's confidentiality, identification of an authorized spokesperson, and planning for structured communication with additional representatives of the media. Efforts to prevent exaggerated estimates of risk to the community can also be discussed.

Although these exercises are presented as a series of scenarios followed by questions, it is suggested that they be used as an opportunity to engage participants in a broader discussion of the issues raised and allow for additional questions to be addressed. For this reason, the leader or leaders of each exercise should prepare by reviewing relevant material in this book regarding the approach to initial management issues, as well as the steps to be taken to prevent transmission of Ebola infection. It may be particularly helpful to have the discussion led by both administrative and clinical personnel.

For the purpose of these scenarios, the premise is that transmission of Ebola infection is currently occurring in the countries of West Africa specified, as it was during the 2014–2016 outbreak.

Although this outbreak has come under control at the time of this writing, and precautions and procedures in many health care facilities may have been relaxed, these exercises are based on experience drawn from the global conditions surrounding the 2014–2016 experience. Similar procedures would be appropriate and/or adaptable in future outbreaks of EVD.

Because of the unpredictable nature of special pathogen outbreaks, simulation drills should be considered in addition to tabletop exercises. In drills, participants actually carry out their functions on people playing the role of patients, such as taking histories, doing physical examinations, imposing isolation, and transporting suspected cases. This tests the physical implementation of procedures relevant to infection control precautions, medical care, and permits rehearsal of key steps. Such simulation drills also permit the identification of physical constraints (such as adequate isolation facilities) that may not be anticipated through tabletop exercises.

It should be noted that hospitals and other health care facilities differ greatly in their ability to effectively provide isolation for patients suspected of having EVD and in their staff's access to and training in the use of personal protective equipment (PPE) and appropriate isolation rooms.

Regional centers for treatment of patients with EVD have been created in the United States (see Chapter 2). However, identifying patients appropriate for these centers, transporting them and, most important, providing the initial care and monitoring required before the transfer, represent logistical challenges.

If the incidence of EVD in the world is low, as it is at the time of this writing, these exercises can remind staff on a regular basis of the challenges posed by the 2014–2016 epidemic of EVD in West Africa. Few cases were seen outside of the countries of West Africa that were dramatically affected: Guinea, Sierra Leone, and Liberia. Nonetheless, because of the highly dangerous nature of EVD and the fact that it could be transmitted through simple contact, a high degree of fear and anxiety gripped much of the world during the epidemic. Exercises such as those illustrated in this chapter can serve to remind health care staff of the complexity and urgency of adequately assessing risk of travelers from areas where transmission of EVD is possible. In addition, training staff in infection control procedures through an Ebola-based tabletop exercise can prepare them for other contagious diseases.

The specific resources available to hospitals and providers to adequately and safely care for patients who might have EVD are under constant review. As the likelihood of a patient with EVD arriving unexpectedly in a hospital or clinic may be low, the actions to be taken in that situation are specific and critical.

What follows are scenarios of specific situations involving likely or possible EVD patients.

CASE 1.   PATIENT ARRIVING FROM WEST AFRICA WITH NO SYMPTOMS OF EBOLA

Intended participants

> *Emergency department greeting and registration staff*
> *Emergency department triage staff*
> *Emergency department nursing staff*
> *Emergency department medical staff*
> *Hospital infection control staff*
> *Hospital security staff*
> *Hospital housekeeping staff*
> *Hospital laboratory staff*

SCENARIO PART I

A 46-year-old Liberian man arrives at JFK Airport in New York City on a flight from Monrovia that had a three-hour stop in Frankfurt. The man, who came to visit relatives in the United States, is referred to the emergency department because he came from Liberia, which is having an Ebola epidemic.

QUESTIONS PART I

What questions should be asked by staff having initial contact?

The role of staff involved in greeting, directing, or triaging the patient is to identify significant travel risk within the 21 days prior to arrival. For this reason, up-to-date information about the countries and regions experiencing transmission of Ebola virus should be readily available and posted clearly.

Since the maximum incubation period of EVD is thought to be approximately 21 days, the date and time of departure from the endemic region should be determined.

(Continued)

CASE 1. (Continued) PATIENT ARRIVING FROM WEST AFRICA WITH NO SYMPTOMS OF EBOLA

What should be done if the man left the endemic area within the previous 21 days?

The patient should be provided with a surgical mask and escorted to an isolation room or other appropriate location where separation from other patients, visitors, and staff can be maintained during further questioning and evaluation.

What potential exposures and additional symptoms should the person be asked about?

The first priority in assessing patients is to confirm whether or not they have symptoms that may represent EVD. The most frequent symptom of EVD is fever. Although a fever of 100.4°C has been suggested as a minimum to warrant the institutions of precautions for EVD, any elevation of temperature should be of concern. If the person reports earlier fever and/or chills during travel, evaluation should proceed as though the person has current fever. Gastrointestinal symptoms, including nausea, vomiting, and diarrhea, are also associated with EVD as are, in some cases, conjunctivitis, headache and, rarely, a generalized rash.

The person should be questioned regarding possible exposures to known or likely cases of EVD. The most concerning exposure would involve likely contact with body fluid as in close physical, including sexual, contact. Such direct contact with a known or suspected case, including contact with a dead body, would justify sustaining isolation procedures until EVD can be definitively excluded. It is important to keep in mind that such exposure may not have been recognized by the patient. For this reason questions should be broad and include inquiries about febrile illnesses among acquaintances, including household contacts who were sick or visits to health care facilities where patients with EVD might have been present.

If no symptoms suggestive of EVD are present, it is nonetheless necessary to screen for potential exposures that occurred in the endemic country.

(Continued)

CASE 1. (Continued)　PATIENT ARRIVING FROM WEST AFRICA WITH NO SYMPTOMS OF EBOLA

What potential exposures that occurred in the endemic country should be identified?

Occupational

The exposures involving close physical contact that most frequently resulted in transmission of Ebola virus infection in the West African outbreak were seen among health care workers and individuals participating in funeral and burial procedures. If the traveler is a health care worker, he should be questioned about potential exposure to patients or other staff known or suspected of having EVD within the past 21 days. An understanding of the facility in which he may have seen patients and the nature of the personal protective equipment used (PPE) would allow for a more accurate estimate of risk. Individuals handling dead bodies of EVD victims either in health care facilities or during burial rituals were particularly at high risk during the 2014–2016 West African outbreaks because of the high concentrations of virus typically present on the skin of the corpses. For this reason, the traveler should be specifically questioned about attendance at funerals in order to assign a risk category.

Nonoccupational

Relatives and others tending to individuals ill with EVD should also be assumed to be at high risk of infection. This would include direct physical contact with the body, living or dead, and exposure to body fluids either indirectly or through contact with bedding, clothing, and so on.

What other questions should be asked about his travel experience?

Has anyone accompanied him?

Clearly anyone traveling with the patient must also be screened in a similar fashion regarding travel and potential symptoms and additional contacts. Friends or relatives who have accompanied the patient to the ED should be included even if they are not expressing concerns regarding their own health.

Did he take malaria prophylaxis?

The early stages of malaria and EVD may be identical. For this reason, it is important to also gauge the risk of malaria, a much more common

(Continued)

CASE 1. (Continued)   PATIENT ARRIVING FROM WEST AFRICA WITH NO SYMPTOMS OF EBOLA

infection that is endemic in the same regions as EVD was during the 2014–2016 outbreak and the areas in which any future naturally occurring outbreaks are likely to arise. It is rare for a traveler native to an endemic area to have taken prophylactic medications to prevent malaria, but the specifics of the individual's itinerary may have caused them to do this. Knowing that appropriate prophylaxis was taken allows the health care provider to better assess that likelihood of malaria. This may be of particular importance in situations where the traveler has symptoms compatible with both malaria and EVD and laboratory facilities needed to distinguish the two infections are not readily available. In such circumstances, treatment for malaria may be necessary prior to a confirmed diagnosis.

Circumstances of his travel

As was learned during the 2014–2016 outbreaks, concerns regarding exposure of other passengers to a patient with symptomatic EVD can reach high levels. Although the simple sharing of a flight, or sitting near someone in a cab, a departure area, or other airport facilities poses no recognized risk of transmission if there was no opportunity for direct exposure to body fluids, the traveler should be questioned about the details of his travel from departure in the endemic country until arrival at his destination. These details, some of which are suggested here, would be of concern only if symptoms had developed during travel since EVD appears to be transmissible only after the onset of symptoms, particularly fever. If likely significant exposure did take place, the process of notifying the airline and public health officials should begin without delay. It should be noted, however, that no such secondary cases were documented among individuals exposed during travel in the West African epidemic. Nonetheless, among the areas to be clarified are

Was he traveling with anyone from the endemic area who subsequently developed symptoms potentially representing EVD?

The possibility of close physical contact with anyone accompanying him should be thoroughly explored. That individual should be identified and evaluated for symptoms of EVD during or after contact with the traveler.

(Continued)

CASE 1. (Continued)   PATIENT ARRIVING FROM WEST AFRICA WITH NO SYMPTOMS OF EBOLA

What did he do during any layover?

If the traveler was symptomatic during the layover, it is necessary to inquire about the details of potential close personal contact during his time either in or out of the airport vicinity in order to identify potential exposures. Of particular interest would be if he underwent a massage, a service offered at some airports, or if, while ill, others physically assisted him.

SCENARIO PART II

The patient denies all symptoms of possible EVD. While in Liberia, however, he did attend the funeral of his four-year-old nephew who had died after several days of fever, chills, and worsening diarrhea. He did not touch the body and was unaware of any exposure to body fluids from the child or anyone else at the funeral.

QUESTIONS PART II

How should the risk of EVD in this patient be categorized?

The likelihood that he has contracted EVD is uncertain. Although his potential exposure seems unimpressive, the fact that he was in proximity to the body of a child who might have died of the disease should be recognized as an indication for close monitoring of his status throughout the possible incubation period of 21 days. He should be instructed according to the policy of local health authorities. The options would be either daily visit by public health workers to evaluate his clinical status and/or twice daily temperature checks by the patient. He should be instructed to contact the individual designated by the health department daily to report his status. He should immediately make contact at any time of the day or night if he develops fever, so that transportation under appropriate precautions to a designated emergency department can be arranged.

As this is being carried out, efforts should be made to establish the cause of death of the child.

(Continued)

CASE 1. (Continued) PATIENT ARRIVING FROM WEST AFRICA WITH NO SYMPTOMS OF EBOLA

Key points:

- Individuals thought to be arriving from endemic areas should be questioned first about their specific travel itinerary, including the time since leaving the endemic country.
- Initial symptoms of EVD are typically nonspecific. Fever is of most concern.
- Provision should be made in the emergency department for individuals with the appropriate travel history and possible symptoms of EVD to be identified promptly and escorted to appropriate isolation areas for further evaluation.
- Anyone accompanying the patient should be questioned in a similar fashion regarding the recent travel and possible close exposure to the patient.
- If symptoms are suspected, the details of his travel and potential exposure of others will be necessary to determine the likelihood of risk and the need for testing and other measures.

CASE 2. PATIENT ARRIVING FROM WEST AFRICA WITH SYMPTOMS OF EBOLA

Intended participants

Emergency department greeting and registration staff
Emergency department triage staff
Emergency department nursing staff
Emergency department medical staff
Hospital infection control staff
Hospital security staff
Hospital housekeeping staff
Hospital laboratory staff

(Continued)

CASE 2. (Continued)   PATIENT ARRIVING FROM WEST AFRICA WITH SYMPTOMS OF EBOLA

SCENARIO PART I

A 22-year-old female college student from the United States, who has been working in Sierra Leone for two months as part of a medical relief effort unrelated to Ebola, comes to the emergency department three days after her return. She is complaining of high fever, chills, and nausea.

QUESTIONS PART I

What steps should the emergency room staff take?

As in the previous exercise, the role of the staff greeting, registering, and triaging the patient is to establish that she has traveled to an endemic country within 21 days. This should be determined rapidly by a process in which all patients are questioned about recent travel, regardless of presenting complaints. Her history, as described above, should prompt immediate concern that she very possibly acquired an infection while in Sierra Leone. Public health authorities should be notified immediately, and anyone accompanying her should be evaluated. Her symptoms are nonspecific but consistent with EVD. For this reason, she should be given a surgical mask and escorted to an area in the emergency department where separation from other patients, visitors, and staff can be maintained while she is questioned and examined. After this is done, staff entering the room to interview and examine her should wear appropriate personal protective equipment. (See CDC recommendations in Chapter 7 and the Appendix.)

What additional information should be obtained?

She should be questioned in detail about her activities in Sierra Leone in order to determine the following:

What exactly were her duties? Did she have contact with patients in a health care setting? If so, in what type of health care setting did the contact occur? Were the patients hospitalized or ambulatory? If hospitalized, were isolation procedures being followed? What were the details of these procedures? If she indicates that she worked in a health care facility, the nature of the patients in that facility should be determined. Did she have contact with anyone known or suspected to have EVD within the recognized incubation period of 21 days? What was the nature of this contact?

(Continued)

CASE 2. (Continued)   PATIENT ARRIVING FROM WEST AFRICA WITH SYMPTOMS OF EBOLA

Did it involve touching patients, their clothing, their bedding, and so on, or contact with body fluids or handling clinical specimens?

Any body fluids, with which the individual had direct, even incidental contact, would potentially be relevant. Any direct physical contact would potentially represent a route of transmission of EVD. Contact with body fluids, such as handling of bed pans, urinals, or intravenous or respiratory equipment would be relevant.

Did she wear personal protective equipment (PPE)? What type?

For protection against transmission of Ebola virus, fully occlusive PPE is recommended. As described in the Appendix, this should include head and face covering, eye covering, a mask and a gown, and boots that leave no unprotected skin or mucous membranes. It is unlikely that equipment meeting this definition and donned and doffed safely would have been used by this person unless she was told she could come into contact with patients with known or possible EVD.

The circumstances under which PPE was used, donned and doffed would be critical to assess the likelihood of transmission of Ebola virus.

Did she have contact with anyone, outside of a health care facility, who was complaining of fever?

If so, what were the circumstances? What was the nature of the contact? If she worked in a health care setting, it might be easy to ignore nonoccupational exposures she might have had in an endemic area. During the 2014–2016 West African epidemics, some of the health care workers who were infected were not working in Ebola treatment facilities and appeared to have contracted the disease outside of their work setting.

Did she take malaria prophylaxis? What medications? What dates?

As noted in the first scenario in this chapter, in evaluating a patient it is important to gauge the risk of malaria, which can be confused with Ebola and is far more common. As a U.S. resident volunteering in West Africa, she is much more likely to have taken prophylaxis than a native West African.

(Continued)

CASE 2. (Continued)   PATIENT ARRIVING FROM WEST AFRICA WITH SYMPTOMS OF EBOLA

SCENARIO PART II

The patient indicates that she did not work in a health care facility, but met with community members in rural villages on five occasions in order to share information about nutrition and sanitary practices as part of a public health campaign designed to lower infant mortality. She was not aware that anyone with whom she met was sick and did not have any close contact, except to shake hands with and embrace a number of people. She lived in a barrack with five other American college students, one of whom had accompanied her throughout her stay and travels and did not report any symptoms to her, although she has not been in touch with that individual since her return. On the final day of her stay, one of the students was complaining of chills and abdominal pain, but she did not know what happened to that person after she left. She said that she and all her fellow American students had taken malaria prophylaxis during their stay in Sierra Leone.

QUESTIONS PART II

Did any of the students with whom she lived have potential contact with EVD?

The travel histories of the students with whom she stayed should be ascertained if possible. The focus should be on their activities and travel within the weeks prior to contact with the patient. Although the maximum incubation period of EVD appears to be approximately 21 days, potential exposure to Ebola patients for twice this interval, that is 42 days, should be considered when clues to secondary transmission are sought. This strategy assumes that contacts of this patient, that is the roommates, may have become ill on the last day of their incubation period, 21 days and that this patient may become ill 21 days after last contact with them.

Did any of the community members she met in her work potentially have EVD?

This may not be the information that is known to her. The communities from which these individuals came may provide evidence of potential exposure to EVD. However, the nature of her physical contact with them

(Continued)

CASE 2. (Continued)   PATIENT ARRIVING FROM WEST AFRICA WITH SYMPTOMS OF EBOLA

as well as the difficult-to-impossible task of obtaining more information about them, likely would lead this to be considered a potential exposure to EVD that cannot be excluded.

How should the risk of EVD in this patient be categorized?

Because the patient has symptoms that are nonspecific but can be considered consistent with EVD, and because she has arrived from a region known to have cases (at the time of this scenario), she should be treated and staff should be protected as though she is at high risk. The fact that she had physical contact with a number of other individuals within the incubation period of EVD, whose current status is unknown, further establishes the high-risk nature of her presentation.

What steps should be taken in her evaluation?

The patient should be placed in an isolation room. It is essential that only trained staff wearing adequate PPE enter her room. Any staff who does not have access to adequate PPE should communicate with her without direct contact and from another room. In facilities designed for EVD isolation, the adjacent room, and the staff and equipment room, is separated from the patient's room by a window and door. In the staff and equipment room, facilities are arranged, so that staff have the equipment and space to don and doff the appropriate PPE under supervision before entering the patient's room.

This patient's presentation, however, is consistent with a number of other infections that could also be life-threatening if not addressed. Among these are malaria and bacterial infections, either localized or disseminated. If the availability of PPE and/or the logistics in adequately employing it are a significant obstacle to do appropriate laboratory tests to exclude other possible serious infections; empiric treatment for these may have to be started as the evaluation for EVD proceeds. Therapy for malaria, dictated by recommended regimens for her countries of travel, typhoid, and other bacterial diseases should be begun if there is any delay in confirming or ruling out these diagnoses. State and local health departments should be contacted for guidance on sending blood specimens to be tested for Ebola as other conditions are being assessed and treated or treated empirically. If a diagnosis of EVD is confirmed or

(Continued)

CASE 2. (Continued)   PATIENT ARRIVING FROM WEST AFRICA WITH SYMPTOMS OF EBOLA

remains suspected, transport of the patient to a regional or national center for treatment should be considered.

Key points:

- In an endemic country, not working in a health care facility does not eliminate all risk.
- In an endemic country, if contacts cannot be established to be noninfected, they should be regarded as possibly infected.
- When symptoms are nonspecific but circumstances raise the possibility of such life-threatening diseases as malaria or typhoid, empiric treatment should be considered even in the absences of laboratory confirmation.
- If 42 days have elapsed since a patient's most recent contact with a possibly infected person, she can be presumed to be uninfected. This reflects the presumed maximum incubation period of 21 days. If she had been exposed on the last day, the 21st day, of an asymptomatic person's incubation period, she would be past the last day of her own potential incubation period after the 42nd day.

CASE 3.   ASYMPTOMATIC HEALTH CARE WORKER RETURNING FROM WEST AFRICA

Intended participants

Emergency department greeting and registration staff
Emergency department triage staff
Emergency department nursing staff
Emergency department medical staff
Hospital infection control staff
Hospital security staff
Hospital housekeeping staff
Hospital laboratory staff

(*Continued*)

CASE 3. (Continued) ASYMPTOMATIC HEALTH CARE WORKER RETURNING FROM WEST AFRICA

SCENARIO PART I

An asymptomatic physician who has been working in an Ebola treatment center in Liberia for six weeks reports to the emergency room for evaluation. He worked in a center in which he always put on full PPE under observation. He was in daily contact with possible or proven patients with EVD. He returned to the United States one week ago. He has come to the emergency room to undergo evaluation for possible EVD, although he has noted no symptoms.

QUESTIONS PART I

What were his duties at the Ebola treatment center?

Potential contact with EVD patients should be assumed. However, the nature and location of the contact may influence the assessment of his risk. For example, if his work was all or in part within a screening facility, he should be asked what precautions were taken with patients who were undergoing this screening prior to their identification as potential EVD patients. Clarification regarding how patients who were undergoing screening were separated from those presumed to have EVD. Were these the same precautions that were taken with patients who were thought to have EVD after screening? Did he wear full PPE during screening and care of patients?

Where and with whom did he live?

Clarification of living arrangements would potentially add insight into possible transmission from other workers. The occupations and duties of those sharing living quarters could point to potential transmission outside of the health care setting. Specific medical information should be sought about any close household contacts, particularly any with illnesses that might have been consistent with EVD.

If any of his household contacts were involved with the Ebola treatment unit, then questions similar to those asked of the patient would be relevant. Since roommates and other household contacts would likely not be available for detailed questioning, it may be best to assume that partial information about other individuals with whom he lived should not be used to gauge the level of risk of this patient unless he was aware that any had symptoms suggestive of EVD.

(Continued)

CASE 3. (Continued)   ASYMPTOMATIC HEALTH CARE WORKER RETURNING FROM WEST AFRICA

What activities did he participate in other than work-related duties?

It is important to question the patient about nonwork-related activities in which he participated and, in particular, whether any such activities brought him into close physical contact with individuals known to have symptoms suggestive of EVD or if any contacts of these individuals had been diagnosed with EVD or suspected EVD. None of his roommates had become infected.

SCENARIO PART II

The doctor said he was involved in treatment of symptomatic Ebola patients daily. He said he was fully trained in the use of fully occlusive PPE, used it without exception, and did not recall any breaches in donning and doffing procedures.

QUESTIONS PART II

How should his risk be categorized?

This individual had contact with patients with EVD during his work in West Africa. The details of his exposure at the treatment center are important in gauging the likelihood of his acquiring infection there. However, even meticulous use of PPE and care in contact with patients cannot be presumed to have been effective in preventing transmission. His contact with roommates and others in the community in which he worked cannot be used to eliminate the possibility that he has contracted EVD even if the details of those contacts do not seem likely to have resulted in transmission.

What course of action should be followed?

This individual, if truly asymptomatic, should be instructed to take his temperature several times each day at home and to report any fever or new suggestive symptoms to public health authorities, so that a full evaluation can be planned and transportation can be arranged to take him to a facility where this can be safely accomplished. It should be recognized that transmission of EVD prior to the onset of symptoms appears to be

(Continued)

CASE 3. (Continued)   ASYMPTOMATIC HEALTH CARE WORKER RETURNING FROM WEST AFRICA

extremely unlikely. However, with the onset of symptoms, person-to-person transmission becomes increasingly likely over the first several days of illness. For this reason, transportation to a prepared health care facility should be carefully arranged through public health authorities to avoid transmission to others.

Key points:

- In a person who has direct contact with Ebola patients, even use of recommended PPE cannot rule out transmission.
- A person who is at risk of Ebola but is asymptomatic does not need to be hospitalized if he can be relied upon to monitor himself and immediately report fever or other symptoms to public health authorities.

CASE 4.   PATIENT PRESENTING IN NON-ED AMBULATORY SETTING

Intended participants

Ambulatory care greeting and registration staff
Ambulatory care clinical staff
Emergency department greeting and registration staff
Emergency department triage staff
Emergency department nursing staff
Emergency department medical staff
Hospital infection control staff
Hospital security staff
Hospital housekeeping staff
Hospital laboratory staff

(Continued)

CASE 4. (Continued)   PATIENT PRESENTING
IN NON-ED AMBULATORY SETTING

SCENARIO PART I

A 35-year-old woman presents to the walk-in area of an ambulatory care site complaining of fever and diarrhea for the past two days. She returned from visiting relatives in Conakry, Guinea five days earlier. She had stayed in Guinea for 10 days and had traveled alone. None of her relatives or anyone else she had visited during her trip had been ill. However, several of them had heard of possible cases of EVD in a local hospital. None of her relatives or others she visited worked in this hospital or had any contact with sick individuals as far as they were aware of during her stay. Her flight back to the United States had a three-hour layover in Amsterdam.

QUESTIONS PART I

How certain was she that any of the family members she visited subsequently did not develop symptoms that may represent EVD?

It should be recognized that the patient may not have full knowledge of the medical status of the family members she visited. Early symptoms, such as low fever, may not have been emphasized. In addition, since she left the family members several days to over one week earlier, she may not have full knowledge of their current medical conditions.

Does she have any knowledge of the current health conditions of nonfamily members with whom she may have had direct or indirect contact?

Information about the medical conditions of others with whom she had contact may be unknown to her. However, inquiring about this may reveal important possible insight into her current symptoms. If any relatives or other associates with whom she had contact during her trip actually had EVD, they are likely to have become severely ill after she last had contact with them. This information may be available to her. It is important to question the patient in detail about these potential contacts and their current clinical status if known to her.

What were her activities during her layover in Amsterdam?

Although the answer may potentially shed light on the nature of her current illness, it is extremely unlikely that she acquired EVD in

(Continued)

CASE 4. (Continued)   PATIENT PRESENTING
IN NON-ED AMBULATORY SETTING

Amsterdam, because it has been exceedingly rare outside of Africa. (During the 2014–2016 outbreaks, which provides the fictional context for these scenarios, the only reported EVD in the Netherlands was in a UN peacekeeper who was medically evacuated from Liberia for treatment.) The likelihood that she transmitted EVD while in Amsterdam is extremely unlikely because she had not yet developed symptoms while there. Even if her symptoms were mild and unrecognized, she was unlikely to have had the degree of physical contact necessary for transmission.

SCENARIO PART II

The patient calls her mother in Conakry from the emergency department and asks about the health of the relatives with whom she met. The mother says she is sure that no one is ill. She says that she spent her layover in Amsterdam at the airport.

QUESTIONS PART II

What is the extent of EVD in the areas she visited in West Africa?

The likelihood that this patient has EVD would be determined by the status of the disease in West Africa or any other region she had visited within the 21 days preceding the onset of her symptoms. Since she did not work in or visit a health care facility and had no contact with anyone who was known to have symptoms at the time, her risk of having EVD at this time would appear to be low. If, however, EVD transmission was prevalent in Conakry at the time she visited, her symptoms and the apparent incubation period of her illness should be considered compatible with EVD and procedures for isolation and diagnosis should be followed.

How should an ambulatory care center handle a
person presumed to be at risk of EVD?

Protection of hospital staff through the isolation of the patient and the use of effective fully occlusive PPE would represent a challenge in an ambulatory care center. The use of this case for illustrative purposes and for discussion would be best to serve as a refresher to ambulatory staff

(Continued)

CASE 4. (Continued)   PATIENT PRESENTING
IN NON-ED AMBULATORY SETTING

about the procedures within their institution to be used to evaluate and transport a patient presumed to be at risk of EVD.

What other diagnoses should be considered?

The symptoms and travel history of this patient raise other possible diagnoses as well, including malaria and typhoid, which would be evaluated according to similar exposure possibilities. EVD is, of course, unique among febrile illnesses in returning travelers because of the substantial risk of spread to health care personnel and the especially rigorous forms of isolation that it requires. The urgency of excluding malaria and typhoid, which she is far more likely to have than EVD, might lead to a decision to treat empirically for them. However, the likelihood of EVD in this case might be remote enough that conditions such as malaria and typhoid could be approached with appropriate lab tests, as they typically are in returning travelers from endemic areas.

Key points:

- Even ambulatory care centers need to train staff to evaluate patients for possible Ebola risk and need to have plans in place for isolation and transport to other facilities.

CASE 5.   PREGNANT WOMAN WITH
POSSIBLE EBOLA EXPOSURE

Intended participants

Emergency department greeting and registration staff
Emergency department triage staff
Emergency department nursing staff
Obstetrical medical staff
Obstetrical nursing staff
Emergency department medical staff
Operating room staff

(Continued)

CASE 5. (Continued)   PREGNANT WOMAN WITH POSSIBLE EBOLA EXPOSURE

Hospital infection control staff
Hospital security staff
Hospital housekeeping staff
Hospital laboratory staff

SCENARIO PART I

A woman who is 28 weeks pregnant arrives in New York City from her native country of Liberia, where she has lived all of her life. She comes to the emergency department directly from the airport because she has been having high fever for the past three days. In Liberia, she had been receiving prenatal care at a hospital in Monrovia, but left when she learned that another woman in her third trimester of pregnancy at that hospital had died of possible EVD approximately one week earlier. She had also heard that the child of that woman died during an emergency delivery.

QUESTIONS PART I

How certain is it that there had been a case of EVD at the hospital in Liberia?

The context for this exercise presumes that EVD transmission is taking place in West Africa. Determining the truth of the patient's account of a pregnant woman and her baby dying of EVD might be possible through contacts with the WHO, CDC, or relief agencies, but that information might not be readily available or confirmable and determining the details would likely be too time-consuming for the medical management of this febrile, pregnant, and returning traveler.

Has this woman had other potential exposures to EVD?

The fact that the patient has attended a prenatal clinic at a hospital in Liberia may be taken as an additional risk factor for exposure. The patient also should be questioned about whether any of her family, friends, or other contacts have had symptoms of EVD or been diagnosed with the disease. If she acquired EVD from a contact in her everyday life, that person is likely to have become extremely ill within the past week, since the average incubation period is eight days and she has been

(Continued)

CASE 5. (Continued)   PREGNANT WOMAN
WITH POSSIBLE EBOLA EXPOSURE

ill for three days. The absence of such illness among her contacts is potentially reassuring information. Yet even so, her potential exposure to other pregnant patients in the hospital clinic would continue to represent a risk of contagion.

SCENARIO PART II

The patient says she knows of no household contacts who have recently become ill. She feels strongly that the pregnant woman who died had Ebola. This belief was so strong that it motivated her to leave the country.

QUESTIONS PART II

How should this patient's care proceed?

Based on the circumstances of this case, EVD cannot be excluded in this patient. As a result, she must be placed in an appropriate isolation and all personnel having direct contact with her should don (under observation) full PPE.

How should decisions proceed about the mother and the fetus?

As reviewed in Chapter 4, the prognosis of EVD in pregnancy is extremely poor, for both mother and child. In addition, symptoms associated with pregnancy, particularly nausea and vomiting, morning sickness, may mimic the symptoms of EVD.

Presuming the decision has been made to proceed as though the woman may have EVD, a decision regarding the viability of the fetus and the possible indication for emergency caesarean section must be made. This complex combination of medical circumstances was not faced in patients evacuated from West Africa to the United States or other Western countries during the 2014–2016 outbreak. In West Africa, the mortality of fetuses or children born to mothers with EVD was essentially 100%. Nonetheless, consideration would have to be given to delivering the child while saving the mother. The viability of the fetus should be assessed and the likely effect on the mother's condition of emergency delivery would have to be considered. Since patients with EVD may rapidly develop physiological and laboratory abnormalities

(Continued)

CASE 5. (Continued)   PREGNANT WOMAN WITH POSSIBLE EBOLA EXPOSURE

that influence the mother's ability to undergo delivery of the fetus, any effort intended to save both the mother and the child must be carefully coordinated and developed quickly. U.S. hospitals that have exercised and prepared for the treatment of patients with EVD do not necessarily have drilled protocols to address this situation. Transfer to a regional Ebola center might be considered if medically and logistically feasible. Obviously, the coordination of medical, obstetrical, and pediatric personnel and expertise would be extremely challenging. The meticulous use of effective personal protective equipment and the strategies to be used to monitor the mother's and child's conditions would require extensive planning.

Key points:

- Ebola is extremely dangerous to a pregnant woman and her developing fetus.
- If Ebola is suspected in a pregnant woman, expertise in complex deliveries should be sought and appropriate PPE will be required for all staff, including obstetrical and pediatric staff involved in the care.
- If clinically feasible, transfer to a regional Ebola treatment center should be strongly considered.

REFERENCES

Ganguli I, Chang Y, Weissman A, Armstrong K, Metlay JP. Ebola risk and preparedness: A national survey of internists. *J Gen Intern Med* 2015. doi:10.1007/s11606-015-3493-1.

Highsmith HY, Cruz AT, Guffey D, Minard CG, Starke JR. Ebola knowledge and attitudes among pediatric providers before the first diagnosed case in the United States. *Pediatr Infect Dis J* 2015;34(8):901–903.

Valerio L, Perez-Quilez O, Roure S et al. When information does not translate into knowledge. Ebola and hemorrhagic fevers knowledge among primary care physicians and nurses. *Open J Prev Med* 2015;5:122–127.

Appendix

INTRODUCTION

Additional detailed information is included in this section:

1. Lessons learned from the 2014–2016 Ebola epidemics.
2. Full CDC instructions for donning and doffing personal protective equipment (PPE) used in the care of persons known or suspected of having Ebola virus disease (EVD).
3. Epidemiologic risk factors to be considered when evaluating a person for Ebola virus infection.
4. The Ebola timeline including information about outbreaks of EVD since the virus was first recognized as a human pathogen in 1976.
5. Information on African health care workers who died of Ebola virus disease during the 2014–2016 epidemics.
6. Persons who were treated for Ebola in Europe or the United States during the 2014–2016 epidemic.
7. Information regarding other special pathogens.

LESSONS OF THE 2014–2016 EBOLA EPIDEMIC

1. *Occurrence in a new region led to a delay in recognition that Ebola virus disease was present in West Africa*: What appeared to be the initial cases of Ebola in the 2014–2016 West African epidemic occurred in December 2013 in a two-year-old child, his relatives, and contacts at his funeral. The next cases, which were recognized as Ebola, occurred in March 2014. This gap of almost three months suggests that the infection was circulating in West Africa before it was recognized. Since Ebola virus disease symptoms are similar to those of infections that are endemic to the region, for example, malaria, it is possible that cases went undiagnosed and that recognition of the beginning of the outbreak was delayed.

2. *The ease of transmission to health care workers was not initially anticipated and preparations to prevent it were not in place*: More than 800 health care workers, including two caring for a patient in the United States, contracted Ebola infection during the 2014–2016 outbreak. This represented an unprecedented occupational risk of infection with this virus. The reasons for this included inadequate personal protective equipment and waste disposal systems and practices, as well as difficulty in determining which patients had Ebola infection.

3. *Public fear of the infection and suspicion about the motives of health care workers interfered with the medical response*: In a variety of countries and settings, there were public fears of contagion and concerns that full and accurate information about the outbreak was not being provided. In some situations in West Africa, these concerns led to violence against health care workers and avoidance of Ebola treatment units. In the United States, government officials in some cases restricted travel of health care workers and insisted on quarantine of returning health care workers.

4. *Health care facilities in Africa were unprepared to diagnose and care for the number of patients*: The three countries primarily involved in the 2014–2016 outbreaks, Guinea, Liberia, and Sierra Leone, were among the poorest in the world. They had inadequate hospital and laboratory facilities to manage the outbreak. A profound shortage of health care workers in all three countries contributed greatly to the crisis.

5. *Fears of contagion among the U.S. public resulted in misguided efforts at control*: Despite the fact that only two cases of transmission of Ebola occurred in the United States and that those were in an intensive care unit, the public perception of risk was great. A public opinion poll in November 2014 revealed that the U.S. public believed that Ebola was the most pressing health care crisis (SteelFisher 2015). Misperceptions involving routes of transmission and the likelihood of becoming infected were common and risk was greatly exaggerated. As public awareness of Ebola as a news story reached approximately 80%, due, in large part, to the frequent coverage of the epidemic on television, mistrust of government recommendations became common. The temporal proximity to the November 2014 midterm elections may have contributed to exaggerated fears as candidates spoke about Ebola.

6. *A coordinated international effort was necessary to bring the epidemic under control*: The international effort to bring the epidemic under control was ultimately responsible for the dramatic reduction in new cases in early 2015. The cost in personnel, equipment, and funding was both unprecedented and unexpected.

7. *The outbreak affected broader public health planning*: Implications for future outbreaks of Ebola and other emerging infectious diseases were examined anew. In the context of the large-scale outbreak of Zika virus, which began in 2015 in South America, the World Health Organization (WHO) declared it to be a public health emergency of international concern (PHEIC) on February 1, 2016, before the link between the Zika virus and birth defects, including microcephaly, was clear. The WHO had been criticized for a perceived delay in declaring the Ebola outbreak to be a PHEIC and this action was regarded as a more nimble response than the agency had taken toward Ebola (Tavernise and McNeil 2016).

REFERENCES

SteelFisher GK, Blendon RJ, Lasala-Blanco N. Ebola in the United States—Public reactions and implications. *N Engl J Med* 2015;373:789–791.

Tavernise S, McNeil DG. Zika virus a global health emergency, W.H.O. says. *New York Times*, February 1, 2016. http://www.nytimes.com/2016/02/02/health/zika-virus-world-health-organization.html?_r=0.

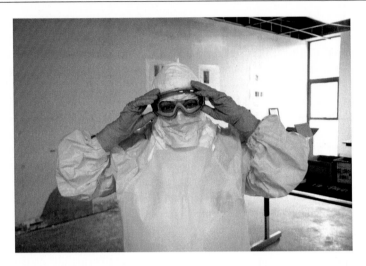

PHOTO A.1 A health care worker wears recommended personal protective equipment during a training course run by the Centers for Disease Control and Prevention in 2014. The three-day course was intended to prepare health providers to work in the West African countries then experiencing an Ebola epidemic. (Courtesy of Cleopatra Adedeji, RRT, BSRT, CDC, Atlanta, GA.)

CENTERS FOR DISEASE CONTROL AND PREVENTION INSTRUCTIONS ON PPE

What follows are the CDC's instructions on PPE use, as of August 27, 2015.

Guidance on Personal Protective Equipment (PPE) to Be Used by Health Care Workers during Management of Patients with Confirmed Ebola or Persons under Investigation (PUIs) for Ebola Who Are Clinically Unstable or Have Bleeding, Vomiting, or Diarrhea in U.S. Hospitals, Including Procedures for Donning and Doffing PPE

Who this is for: Health care workers, supervisors, and administrators at U.S. hospitals.

What this is for: To protect health care workers and other patients at facilities that provide care to a patient with confirmed Ebola or PUI who is clinically unstable or has bleeding, vomiting, or diarrhea by describing protocols for using PPE.

How to use: Incorporate into infection control and safety training for health care workers who provide care to patients with Ebola and use in planning for staffing and supply management.

How it relates to other guidance documents: There are two PPE guidance documents for U.S. hospital workers who may evaluate or care for Ebola patients. Workers should wear this recommended PPE ensemble when evaluating and caring for

1. A person who meets the definition of a person under investigation for Ebola and is
   a. Exhibiting obvious bleeding, vomiting, or diarrhea; OR
   b. Clinically unstable and/or will require invasive or aerosol-generating procedures (e.g., intubation, suctioning, and active resuscitation).
2. A person with confirmed Ebola.

Refer to U.S. health care settings: Donning and doffing personal protective equipment for evaluating persons under investigation for Ebola who are clinically stable and do not have bleeding, vomiting, or diarrhea recommended when evaluating and caring for a PUI who is

1. Not exhibiting obvious bleeding, vomiting, or diarrhea.
2. Clinically stable and will not require invasive or aerosol-generating procedures (e.g., intubation, suctioning, and active resuscitation).

Key Points

- Health care workers caring for patients with Ebola must have received comprehensive training and demonstrated competency in performing Ebola-related infection control practices and procedures.
- PPE that covers the clothing and skin and completely protects mucous membranes is required when caring for patients with Ebola.
- Personnel providing care to patients with Ebola must be supervised by an on-site manager at all times, and a trained observer must supervise each step of every PPE donning/doffing procedure to ensure established PPE protocols are correctly completed.
- Individuals unable or unwilling to adhere to infection control and PPE use procedures should not provide care for patients with Ebola.

Updates to Previous Versions of this Guidance

This Ebola PPE guidance has been updated to add detail, clarify where needed, and improve the format. Specifically, the guidance was updated to

- Expand the rationale for respiratory protection.
- Clarify that the trained observer should not serve as an assistant for doffing PPE.
- Suggest that a designated doffing assistant or *buddy* might be helpful, especially in doffing with the powered air purifying respirator (PAPR) option.
- Modify the PAPR doffing procedure to make the steps clearer.
- Change the order of boot cover removal. Boot covers should now be removed after the gown or coverall.
- Clarify the types of gowns and coveralls that are recommended and provide a link to considerations for gown and coverall selection.
- Emphasize the importance of frequent cleaning of the floor and work surfaces in the doffing area.

Introduction

The following guidance on the types of PPE to be used and the processes for donning (putting on) and doffing (removing) PPE is for all personnel entering the room of a patient hospitalized with Ebola. This guidance reflects lessons learned from the recent experiences of U.S. hospitals caring for patients with Ebola and emphasizes the importance of *training, practice, competence*, and *observation* of health care workers, especially in the correct donning and doffing of PPE.

In health care settings, Ebola is spread through direct contact with blood or body fluids of a person who is sick with Ebola or with objects (e.g., bathroom surfaces and medical equipment) that have been contaminated with infectious blood or body fluids. The virus in blood and body fluids can enter a person's body through broken skin or unprotected mucous membranes, for example, in the eyes, nose, or mouth. For all health care workers caring for patients with Ebola, PPE that fully covers skin and clothing and prevents any exposure of the eyes, nose, and mouth is recommended to reduce the risk of accidental self-contamination of mucous membranes or broken skin. All PPE must be used in the context of a comprehensive infection control program that follows CDC recommendations and applicable Occupational Safety and Health (OSHA) Act of 1970 requirements, including the Bloodborne Pathogens (29 CFR 1910.1030), PPE (29 CFR 1910.132), and Respiratory Protection (20 CFR 1910.134) standards, and other requirements under OSHA (e.g., the General Duty Clause, Section 5(a)(1); and prohibitions against discrimination or retaliation against workers, Section 11(c)).

To protect health care workers who are caring for patients with Ebola, health care facilities must provide on-site management and oversight of adherence to safely using PPE, and implement administrative and environmental controls with continuous safety checks through direct observation of health care workers, including during the PPE donning and doffing steps.

Section 1. Recommended administrative and environmental controls for health care facilities

Protecting health care workers and preventing spread of Ebola to other patients requires that proper administrative procedures and safe work practices be carried out in appropriate physical settings. These include the following:

- At an administrative level, the facility's infection prevention management team (i.e., infection control), in collaboration with the facility's occupational health department and other clinical departments, should

- Establish and implement triage protocols to effectively and promptly identify patients who could have Ebola.
- Designate site managers who are responsible for overseeing the implementation of routine and additional precautions for health care worker and patient safety. These site managers should have experience in implementing protocols for employee safety, infection control, and patient safety. A site manager's sole responsibility is to ensure the safe delivery of clinical care to patients with Ebola. They are responsible for all aspects of Ebola infection control, including access to supplies and ongoing evaluation of safe practices with direct observation of care before, during, and after staff enter an isolation and treatment area.
  - At least one site manager should be on-site at all times in the location where a patient with Ebola is receiving care.
  - Consider engaging the hospital incident command structure to further facilitate implementing Ebola-specific precautions.
- Identify, ahead of time, critical patient care functions and essential health care workers to care for patients with Ebola, collect laboratory specimens, and manage the environment and waste.
- Ensure health care workers have been trained and evaluated in all recommended protocols to safely care for patients with Ebola before they enter the patient care area.
- Ensure that workplace safety programs are in place and have been followed, in particular for OSHA's Bloodborne Pathogens, PPE, and Respiratory Protection standards described above. Coordinate with safety program administrators to ensure that all PPE, including respirators, has been selected on the basis of a written risk assessment and that requirements for medical surveillance, medical clearance, fit testing, training, maintenance, storage, reporting, and so on, are in place for all workers with potential exposure to Ebola.
- Train health care workers on all PPE recommended in the facility's protocols. Health care workers should practice donning and doffing procedures and must demonstrate competency through testing and assessment before caring for patients with Ebola.
- Health care workers should practice simulated patient care activities while wearing the PPE to understand the types of physical stress that might be involved and determine tolerable shift lengths.

- Use trained observers to make certain that PPE is being used correctly and that donning and doffing PPE protocols are being adhered to by using a checklist for each step of the donning and doffing procedure.
- Personnel who are unable to correctly use PPE and adhere to protocols should not provide care for patients with Ebola.

• Document training of observers and health care workers for proficiency and competency in donning and doffing PPE and in performing all necessary care-related duties while wearing PPE.

• Designate spaces so that PPE can be donned and doffed in separate areas to prevent any cross-contamination.

• Key safe work practices include the following:

  • Identify and promptly isolate the patient with Ebola in a single patient room with a closed door and a private bathroom or covered bedside commode.

  • Limit room entry to only those health care workers essential to the patient's care and restrict nonessential personnel and visitors from the patient care area.

  • Monitor the patient care area at all times, and, at a minimum log entry and exit of all health care workers who enter the room of a patient with Ebola.

  • Be able to safely conduct routine patient care activities (e.g., obtaining vital signs and conducting clinically-appropriate examinations, collecting and appropriately packaging laboratory specimens).

  • Dedicate a trained observer to watch closely and provide coaching for each donning and doffing procedure to ensure adherence to donning and doffing protocols.

  • Ensure that health care workers take sufficient time to don and doff PPE slowly and correctly without distraction.

  • Reinforce the need to keep hands away from the face during any patient care and to limit touching surfaces and body fluids.

  • Frequently disinfect gloved hands by using an alcohol-based hand rub (ABHR), particularly after contact with body fluids.

  • Prevent needlestick and sharps injuries by adhering to correct sharps handling practices.

    - Avoid unnecessary procedures involving sharps.
    - Use needleless IV systems whenever possible.

- Immediately clean and disinfect any visibly contaminated PPE surfaces, equipment, or patient care area surfaces using an EPA-registered disinfectant wipe.*
- Regularly clean and disinfect surfaces in the patient care area, even in the absence of visible contamination.
  - Only nurses or physicians should clean and disinfect surfaces in the patient care areas to limit the number of additional health care workers who enter the room.
- Observe (by the site manager or his/her designee) health care workers in the patient room if possible (e.g., through a glass-walled intensive care unit [ICU] room, video link) to identify any unrecognized lapses or near misses in safe care.
- Establish a facility exposure management plan that addresses decontamination and follow-up of health care workers in the case of any unprotected exposure. Training and follow-up should be part of the health care worker training.

Section 2. Principles of PPE

Health care workers must follow the below-mentioned basic principles to ensure that no infectious material reaches unprotected skin or mucous membranes while providing patient care:

- Donning
  - PPE must be donned correctly in proper order before entry into the patient care area; PPE should not be later modified while in the patient care area. The donning activities must be directly observed by a trained observer.
- During patient care
  - PPE must remain in place and be worn correctly for the duration of work in potentially contaminated areas. PPE should not be adjusted during patient care. In the event of a significant splash, the health care worker should immediately move to the doffing area to remove PPE. The one exception is that visibly contaminated outer gloves can be changed while in the patient room and patient care can continue. Contaminated outer gloves can be disposed of in the patient room with other Ebola-associated waste.

---

* EPA-registered disinfectant wipe: Use a disposable wipe impregnated with a U.S. Environmental Protection Agency (EPA)-registered hospital disinfectant with a label claim for a nonenveloped virus (e.g., *norovirus, rotavirus, adenovirus, poliovirus*); see EPA list of Disinfectants for Use against Ebola Virus at http://www.epa.gov/oppad001/list-l-ebola-virus.html.

- Health care workers should perform frequent disinfection of gloved hands using an ABHR, particularly after contact with body fluids.
- If during patient care any breach in PPE occurs (e.g., a tear develops in an outer glove, a needlestick occurs, or a glove separates from the sleeve), the health care worker must move immediately to the doffing area to assess the exposure. The facility exposure management plan should be implemented; including correct supervised doffing and appropriate occupational health follow-up, if indicated by assessment. In the event of a potential exposure, bloodborne pathogen exposure procedures must be followed in accordance with the OSHA Bloodborne Pathogens standards.
- Doffing
  - Removing used PPE is a high-risk process that requires a structured procedure, a trained observer, a doffing assistant in some situations, and a designated area for removal to ensure protection.
  - PPE must be removed slowly and deliberately in the correct sequence to reduce the possibility of self-contamination or other exposure to Ebola.
  - A stepwise process should be developed and used during training and patient care.

Double gloving provides an easy way to remove gross contamination by changing an outer glove during patient care and when removing PPE. Beyond this, more layers of PPE may make it more difficult to perform patient care duties and put health care workers at greater risk for percutaneous injury (e.g., needlesticks), self-contamination during care or doffing, or other exposures to Ebola. If health care facilities decide to add additional PPE or modify this PPE guidance, they must consider the risk/benefit of any modification and train health care workers on how to correctly don and doff for the modified procedure. Donning and doffing steps may need to be adapted on the basis of the specific PPE that is purchased by the hospital. If adaptations are made, facilities must select PPE that offers a similar or higher level of protection than what is recommended here, train health care workers in its use, and ensure they demonstrate competence in its use before caring for a patient with Ebola.

Section 3. Training on correct use of PPE

Training ensures that health care workers are knowledgeable and proficient in donning and doffing PPE before caring for a patient with Ebola. Comfort and proficiency when donning and doffing are only achieved by repeatedly practicing correct use of PPE. Health care workers should be required to demonstrate competency in using PPE, including donning and doffing while being observed by a trained observer, before working with patients with Ebola. Training should be tailored to the intended audience and effectively transmit the required information. In addition, during practice, health care workers and their trainers should assess proficiency and comfort with performing required duties while wearing PPE. People unwilling or unable to fulfill these requirements should not care for a patient with Ebola.

- The following elements are essential for PPE training:
  - How to safely don, adjust, use, and doff the specific PPE that the health care worker will use.
  - How to safely conduct routine clinical care.
  - Limitations of the PPE (e.g., duration of use, degree of protection).
  - What to do in the case of an equipment failure or detection of a breach in PPE.
  - How to maintain PPE and appropriately dispose of it after use.
  - The possible physiologic strain associated with using PPE, and how to recognize and report early signs and symptoms, such as fatigue.
- Training must be interactive and should allow frontline health care workers to practice donning, adjusting, using, and doffing the specific PPE that the employee will use.
- Hospitals should ensure that the trained employees understand the content of the training and can correctly perform the required tasks.
- Hospitals should also ensure that employees can demonstrate how to properly don, use, and doff the same type/model of PPE and respirators that they will use when caring for a patient.
- Regular refresher trainings are essential to maintaining these skills.

Section 4. Use of a trained observer

Because the sequence and actions involved in each donning and doffing step are critical to avoid exposure, a trained observer should read aloud to the health care worker each step in the procedure checklist and visually confirm and document that the step has been completed correctly. The trained observer has the sole responsibility of ensuring that donning and doffing processes are adhered to. The trained observer must be knowledgeable about all PPE recommended in the facility's protocol and the correct donning and doffing procedures, including how to dispose of used PPE, and must be qualified to provide guidance and recommendations to the health care worker. The trained observer will coach, monitor, and document successful donning and doffing procedures, and provide immediate corrective instruction if the health care worker is not following the recommended steps. However, the trained observer should NOT provide physical assistance during doffing, which would require direct contact with potentially contaminated PPE. The trained observer is required to wear PPE, nonetheless, because the coaching role will necessitate being present in the PPE removal area during the doffing process. PPE for the trained observer is described in Section 8. The trained observer should know the exposure management plan in the event of an unintentional break in procedure. A designated doffing assistant or *buddy* might be helpful in some circumstances, for example, during the doffing of the PAPR.

Section 5. Designating areas for PPE donning and doffing

- Ensure that areas for donning and doffing are designated as separate from the patient care area (e.g., patient's room) and that there is a predominantly one-way flow from the donning area to the patient care area to the doffing area.
- Confirm that the doffing area is large enough to allow freedom of movement for safe doffing as well as space for a waste receptacle, a new glove supply, and ABHR used during the doffing process. If using a PAPR with external belt-mounted blower, confirm that there is an area or container designated for collecting PAPR components for cleaning and disinfection, as well as routine maintenance.

Facilities should ensure that space and layout allow for clear separation between clean and contaminated areas. Separate the space into distinct areas and establish a directional, one-way flow of care moving from clean areas (e.g., area where PPE is donned and unused equipment is stored) to the patient room and to the PPE removal area (area where potentially contaminated PPE is removed and discarded). The direction of flow should be marked (e.g., signs on the floor) with visible signage; temporary plastic enclosures can be added if necessary. Existing anterooms to patient rooms have been used for doffing but in many cases are not ideal because of their small dimensions. As an alternative, some steps of the PPE removal process may be performed in a clearly designated area of the patient's room near the door, provided these steps can be seen and supervised by a trained observer (e.g., through a window) and provided that the health care worker doffing PPE can hear the instructions of the trained observer.

Whenever possible, close the end of the hallway of a ward or ICU to through traffic, thereby restricting access to the patient's room to essential personnel who are properly trained in recommended infection prevention practices for caring for patients with Ebola. Designate two adjacent rooms, located on either side of the patient's room, to be cleared of equipment and furniture and used as donning and doffing areas. Glass-enclosed rooms or other designs (e.g., wide glass doors, windows, and video monitoring) to observe ongoing care in the patient room and activity in the doffing area are preferred. The path from the room of the patient with Ebola to an external doffing room should be as short as possible and clearly defined and/or enclosed as a contaminated area that is cleaned frequently along with the doffing area. If areas are reconfigured, the facility should make certain that the space remains compliant with all applicable building and fire codes.

Post signage to highlight key aspects of PPE donning and doffing, including

- Designating clean areas versus contaminated areas.
- Reminding health care workers to wait for a trained observer before removing PPE.
- Listing each step of the doffing procedure.
- Reinforcing the need for slow and deliberate removal of PPE to prevent self-contamination.
- Reminding health care workers to disinfect gloved hands in between steps of the doffing procedure, as indicated below.

Designate the following areas with appropriate signage:

1. *PPE storage and donning area*: This is a clean area outside the patient room (e.g., a nearby vacant patient room, a marked area in the hallway outside the patient room) where clean PPE is stored and where health care workers don PPE before entering the contaminated area and the patient's room. Do not store potentially contaminated equipment (e.g., PAPR components that have not been cleaned and disinfected), used PPE, or waste removed from the patient's room in the clean area. If waste must pass through this area, it must be properly contained.

2. *Patient room*: Use a single-patient room, preferably with a private bathroom; a covered bedside commode with bagging of human waste is an alternative approach. Plan ahead for the need to store many bags of regulated medical waste before their secondary containment. Additional guidance on waste management can be accessed at Ebola-Associated Waste Management and www.osha.

gov/Publications/OSHA_FS-3766.pdf. The door to the patient room should be kept closed. Any item or health care worker exiting this room should be considered contaminated.

3. *PPE doffing area*: Designate an area near the patient's room (e.g., anteroom or adjacent vacant patient room that is separate from the clean area) where health care workers leaving the patient's room can stand to doff and discard their PPE. Alternatively, some steps of the PPE removal process may be performed in a clearly designated area of the patient's room near the door, provided these steps can be seen and supervised by a trained observer (e.g., through a window and provided that the health care worker doffing PPE can hear the instructions of the trained observer). Do not use this designated area within the patient room for any other purpose. Stock gloves in a clean section of the PPE removal area are accessible to the health care worker while doffing.

In the PPE removal area, provide supplies to disinfect PPE and perform hand hygiene and space to remove PPE, including an easily cleaned and disinfected seat where health care workers can remove boot or shoe covers. If space allows, designate stations around the perimeter of the doffing room where each piece of PPE will be removed, moving from more contaminated to less contaminated areas of the room as PPE is doffed. Provide leakproof disposable infectious waste containers for discarding used PPE. Provide a container to collect all reusable PAPR components. Frequently clean and disinfect the PPE removal area, including after each doffing procedure has been completed. One way such cleaning may be achieved is by having another health care worker who has just donned their full PPE clean the doffing area, moving from cleaner to dirtier areas within the doffing area, before entering the patient's room.

Facilities should consider making showers available for use for the comfort of health care workers after doffing PPE at the end of their shift; the heat from wearing PPE is likely to cause significant perspiration.

Section 6. Selecting PPE for health care workers who care for patients with Ebola

This section outlines several PPE combinations and how they should be worn. The key to safely wearing PPE is consistent and correct use reinforced by repeated training and practice. Variations in PPE used to care for patients with Ebola should be avoided within a facility. A facility should select and standardize the PPE to be used by all health care workers who are directly interacting with patients with Ebola. OSHA's Bloodborne Pathogens standard requires employers to establish a written exposure control plan designed to eliminate or minimize employee exposures and should include procedures for donning and doffing the PPE ensemble that is chosen. The protocol must be reviewed by staffs who participate in Ebola care and the trained observer should ensure the protocol is adhered to.

Airborne transmission of Ebola has not been documented in hospitals or households during any of the human outbreaks investigated to date. However, certain procedures (e.g., bronchoscopy and endotracheal intubation) might create mechanically generated aerosols that could be infectious. Such aerosol-generating procedures require additional precautions. Experience in the care of patients hospitalized with Ebola in the United States indicates that the level of care may change unexpectedly and could require an aerosol-generating procedure. Because there might not be time for staff to leave the room to don proper PPE for an aerosol-generating procedure, CDC recommends that all health care workers entering the room of a patient with Ebola wear respiratory protection that would protect them during an aerosol-generating procedure. This would include a NIOSH-certified, fit-tested N-95, or higher respirator, or a PAPR.

Safety and comfort are both critical for health care workers wearing PPE while caring for patients with Ebola. Standardized attire under PPE (e.g., surgical scrubs or disposable garments and dedicated washable footwear) helps the donning and doffing process and eliminates concerns of contaminating personal clothing. Footwear should be closed-toe, soft-soled, washable, and have a closed back. If facilities elect to use different PPE from what is outlined below (e.g., coveralls with either an integrated hood or a surgical hood with integrated full-face shield), they must train health care workers on how to use each type of PPE type and ensure that donning and doffing procedures are adjusted and practiced accordingly. Extra layers of PPE are not advised because they can reduce comfort, field of vision, and mobility, and increase the risk of error and injury while adding no meaningful protection for the wearer.

In this guidance, *impermeable* gowns and coveralls indicates that the material and construction have demonstrated resistance to synthetic blood and simulated bloodborne pathogens. In contrast, fluid-resistant indicates a gown that has demonstrated resistance to water or a coverall that has demonstrated resistance to water or synthetic blood. These categories reflect the currently available U.S. product specifications; specific test methods that assess resistance for these products are listed in Table A.1. When purchasing gowns and coveralls, facilities should follow specifications in Table A.1 to ensure they select gowns and coveralls as described in Sections 5 and 6.

For more details, refer to technical document "Considerations for Selecting Protective Clothing Used in Healthcare for Protection against Microorganisms in Blood and Body Fluids," which provides a more detailed explanation of the scientific evidence and national and international standards, test methods, and specifications for fluid-resistant and impermeable protective clothing used in health care.

TABLE A.1   Specifications for impermeable and fluid-resistant gowns and coveralls

	GOWN	*COVERALL*
Impermeable	Surgical or isolation[a] gown that passes • ANSI/AAMI PB70 Level 4 requirements	Coverall[a] made with fabric and seams/closures that passes • ASTM F1671 (13.8 kPa) or • ISO 16604 ≥14 kPa
Fluid-resistant	Surgical or isolation[a] gown that passes • ANSI/AAMI PB70 Level 3 requirements or • EN 13795 high performance surgical gown requirements	Coverall[a] made of fabric that passes • AATCC 42 ≤1 g and AATCC 127 ≥50 cm H_2O or EN 20811 ≥50 cm H_2O or • ASTM F1670 (13.8 kPa) or • ISO 16603 ≥3.5 kPa

[a]   Testing by an ISO 17025 certified third-party laboratory is recommended.

Section 7. Recommended PPE when caring for a patient with confirmed Ebola or unstable PUI

* *Impermeable garments*:
    * *Single-use (disposable) impermeable gown* extending to at least mid-calf.
    * *Single-use (disposable) impermeable coverall*: Coveralls without integrated hoods are preferred; coveralls with or without integrated socks are acceptable. Coveralls and gowns should be available in appropriate sizes, so people with long arms are able to cover their forearms without gaps between gloves and sleeves when extending their arms to perform normal duties. Consider selecting gowns or coveralls with thumb hooks to the secure sleeves over the inner glove. Facilities that choose to tape gloves will need to ensure that the tape does not tear the gloves or gown/coverall during doffing and that sharp implements, such as scissors, are not needed to remove the tape. Experience in some facilities suggests that taping can increase risk by making the doffing process more difficult and cumbersome; however, other facilities have identified ways to optimize the use of tape and other adherent materials to anchor sleeves over inner gloves. *Scissors should never be used to remove tape or any other part of PPE.*

- *Respiratory protection*: Either a PAPR or disposable, NIOSH-certified N95 respirator should be worn in case a potentially aerosol-generating procedure needs to be performed emergently. PAPRs with a full-face covering and head-shroud make accidental self-contamination during care more difficult (e.g., while adjusting eyeglasses); disposable N95 face piece respirators are less cumbersome and can be easier to doff safely. Any respirator must be used in the context of a comprehensive, written respiratory protection program as required under OSHA Respiratory Protection standard, 29 CFR 1910.134. This standard includes a hazard assessment to ensure appropriate respirator protection, fit testing, medical evaluation, and training of the worker. When required in the occupational setting, tight-fitting respirators cannot be used by people with facial hair that interferes with the face seal.
  - *PAPR*: A hooded respirator with a full-face shield, helmet, or headpiece. Any reusable helmet or headpiece must be covered with a single-use (disposable) hood that extends to the shoulders and fully covers the neck and is compatible with the selected PAPR. If a hood is used over the PAPR, it must not interfere with the function of the PAPR. The facility should follow manufacturer's instructions for decontaminating reusable components and, on the basis of those instructions, develop facility protocols that include designating responsible personnel who ensure that the equipment is safely and appropriately reprocessed and that batteries are fully charged before reuse.
    - A PAPR with a self-contained filter and blower unit integrated inside the helmet can facilitate doffing.
    - A PAPR with external belt-mounted blower unit requires an additional doffing step, as described below.
  - *N95 respirator*: Single-use (disposable) N95 respirator or higher in combination with single-use (disposable) surgical hood extending to shoulders and single-use (disposable) full-face shield. If N95 respirators are used instead of PAPRs, health care workers should be carefully observed to ensure that they do not inadvertently touch their faces under the face shield during patient care.
- *Single-use (disposable) examination gloves with extended cuffs*: Two pairs of gloves should be worn, so that a heavily soiled outer glove can be safely removed and replaced during care. At a minimum, outer gloves should have extended cuffs. Double gloving also allows potentially contaminated outer gloves to be removed during doffing to avoid self-contamination.

- *Single-use (disposable) boot covers* that extend to at least mid-calf. In addition, single-use (disposable) ankle-high shoe covers (*surgical booties*) worn over boot covers may be considered to facilitate the doffing process, reducing contamination of the floor in the doffing area thereby reducing contamination of underlying shoes. Although the use of shoe covers over boot covers may be analogous to using double gloves to ensure safe doffing, the risk of significant contamination to underlying shoes from the floor during the doffing process is very low relative to the risk of gloved hand contamination. Thus facilities may consider methods other than shoe covers worn over boot covers to facilitate doffing of footwear including, most importantly, frequent cleaning of the floor in the doffing area. Boot and shoe covers (if the latter are used) should allow for ease of movement and must not present a slip hazard to the wearer.
  - *Single-use (disposable) shoe covers* are acceptable only if they will be used in combination with a coverall with integrated socks.
- *Single-use (disposable) apron* that covers the torso to the level of the mid-calf should be used over the gown or coveralls if patients with Ebola are vomiting or have diarrhea, and should be used routinely if the facility is using a coverall that has an exposed, unprotected zipper in the front. An apron provides additional protection, reducing the contamination of gowns or coveralls by body fluids, and providing a way to quickly remove a soiled outer layer during patient care. Select an apron with a neck strap that can be easily broken or untied to avoid having to pull the strap over the head, which makes it easier to remove without self-contamination when exchanging a soiled apron during care, or when removing the apron during the doffing procedure.

Section 8. Recommended PPE for trained observer and doffing assistant during observations of PPE doffing

The trained observer should not enter the room of a patient with Ebola but must be in the PPE donning and doffing area to observe donning and doffing procedures, as outlined in Section 7. The following PPE are recommended for trained observers and doffing assistants observing the doffing process:

- Single-use (disposable) fluid-resistant gown that extends to at least mid-calf or single-use (disposable) fluid-resistant coverall without integrated hood.
- Single-use (disposable) full-face shield.

- Single-use (disposable) surgical mask.
- Single-use (disposable) gloves with extended cuffs. Two pairs of gloves should be worn. At a minimum, outer gloves should have extended cuffs.
- Single-use (disposable) ankle-high shoe covers. Shoe covers should allow for ease of movement and should not present a slip hazard to the wearer.

Trained observers should don and doff selected PPE according to the same procedures as outlined below.

Facilities may elect to use impermeable gowns or coveralls for their trained observers to standardize the PPE in the unit, for ease of training personnel on a single item, and to prevent health care personnel entering the patient care area from inadvertently selecting a fluid-resistant gown or coverall instead of the recommended impermeable garment. If facilities elect to use fluid-resistant gowns or coveralls for their trained observers, they must take measures (e.g., staff training, good signage, clear labeling of the product, and good inventory management practices) to ensure that the correct garment is selected by appropriate personnel.

Section 9. Recommended sequences for donning PPE

Section 9A. Donning PPE, PAPR option
This donning procedure assumes that the facility has elected to use PAPRs. An established protocol facilitates training and compliance. A trained observer should verify compliance with the protocol.

1. *Engage trained observer:* The donning process is guided and supervised by a trained observer, who confirms visually that all PPE is serviceable and has been donned successfully. The trained observer should use a written checklist to guide and confirm each step in donning PPE and can verify the integrity of the ensemble. No exposed clothing, skin, or hair of the health care worker should be visible at the conclusion of the donning process.
2. *Remove personal clothing and items:* Change to surgical scrubs (or disposable garments) and dedicated washable (plastic or rubber) footwear in a suitable clean area. No personal items (e.g., jewelry including rings, watches, cell phones, pagers, and pens) should be brought into the patient room. Long hair should be tied back. Eye glasses should be secured with a tie.
3. *Inspect PPE before donning:* Visually inspect the PPE ensemble to be worn to ensure that it is in a serviceable condition, all required

PPE and supplies are available, and the sizes selected are correct for the health care worker. The trained observer should review the donning sequence with the health care worker before the donning process and read it aloud to the health care worker in a step-by-step fashion.

4. *Put on boot covers*: If a coverall without integrated socks is worn, the upper band of the boot cover will be worn UNDER the pants leg of the coverall to prevent pooling of liquids between the coverall pants leg and upper band of boot cover. This step can be omitted if wearing a coverall with integrated socks.

5. *Put on inner gloves*: Put on first pair of gloves.

6. *Put on gown* or *coverall*: Put on gown or coverall. Ensure gown or coverall is large enough to allow unrestricted freedom of movement. Ensure cuffs of inner gloves are tucked under the sleeve of the gown *or* coverall.

   a. If a PAPR with a self-contained filter and blower unit that is integrated inside the helmet is used, then the belt and battery unit must be put on before donning the impermeable gown *or* coverall, so that the belt and battery unit are contained under the gown *or* coverall.

   b. If a PAPR with external belt-mounted blower is used, then the blower and tubing must be on the outside of gown *or* coverall to ensure proper airflow.

7. *Put on outer gloves*: Put on second pair of gloves (with extended cuffs). Ensure the cuffs are pulled over the sleeves of the gown *or* coverall.

8. *Put on respirator*: Put on PAPR with a full-face shield, helmet, or headpiece.

   a. If a PAPR with a self-contained filter and blower unit integrated inside the helmet is used, then a single-use (disposable) hood that extends to the shoulders and fully covers the neck must also be used. Be sure that the hood covers all of the hair and the ears, and that it extends past the neck to the shoulders.

   b. If a PAPR with external belt-mounted blower unit and attached reusable headpiece is used, then a single-use (disposable) hood that extends to the shoulders and fully covers the neck must also be used. Ensure that the hood covers all of the hair and the ears and it extends past the neck to the shoulders.

9. *Put on outer apron (if used)*: Put on a disposable apron to provide an additional layer for the front of the body.

10. *Verify*: After completing the donning process, the trained observer should verify the integrity of the ensemble. The health care worker

should be able to extend the arms, bend at the waist, and go through a range of motion sufficient for patient care delivery while all remaining correctly covered. A mirror in the room can be useful for the health care worker while donning PPE.

Section 9B. Donning PPE, N95 respirator option
This donning procedure assumes that the facility has elected to use N95 respirators. An established protocol facilitates training and compliance. Use a trained observer to verify successful compliance with the protocol.

1. *Engage trained observer*: The donning process is guided and supervised by a trained observer who confirms visually that all PPE is serviceable and has been donned successfully. The trained observer should use a written checklist to confirm each step in donning PPE and verify the integrity of the ensemble. No exposed clothing, skin, or hair of the health care worker should be visible at the end of the donning process.
2. *Remove personal clothing and items*: Change into surgical scrubs (or disposable garments) and dedicated washable (plastic or rubber) footwear in a suitable, clean area. No personal items (e.g., jewelry including rings, watches, cell phones, pagers, and pens) should be brought into patient room. Long hair should be tied back. Eye glasses should be secured with a tie.
3. *Inspect PPE before donning*: Visually inspect the PPE ensemble to be worn to ensure it is in serviceable condition, all required PPE and supplies are available, and the sizes selected are correct for the health care worker. The trained observer should review the donning sequence with the health care worker before donning begins and read it aloud during donning in a step-by-step fashion.
4. *Put on boot covers*: If a coverall without integrated socks is worn, the upper band of the boot cover will be worn UNDER the pants leg of the coverall to prevent pooling of liquids between the coverall pants leg and upper band of boot cover. This step can be omitted if wearing a coverall with integrated socks.
5. *Put on inner gloves*: Put on first pair of gloves.
6. *Put on gown or coverall*: Put on gown *or* coverall. Ensure gown *or* coverall is large enough to allow unrestricted freedom of movement. Ensure cuffs of inner gloves are tucked under the sleeve of the gown *or* coverall.
7. *Put on N95 respirator*: Put on N95 respirator. Complete a user seal check.

8. *Put on surgical hood*: Over the N95 respirator, place a surgical hood that covers all of the hair and the ears, and extends past the neck to the shoulders. Ensure that hood completely covers the ears and neck.

9. *Put on outer apron (if used)*: Put on a disposable apron to provide an additional layer for the front of the body.

10. *Put on outer gloves*: Put on second pair of gloves (with extended cuffs). Ensure the cuffs are pulled over the sleeves of the gown *or* coverall.

11. *Put on face shield*: Put on full-face shield over the N95 respirator and surgical hood to protect the eyes, as well as front and sides of the face.

12. *Verify*: After completing the donning process, the trained observer should verify the integrity of the ensemble. The health care worker should be able to extend the arms, bend at the waist, and go through a range of motion sufficient for patient care delivery while all remaining correctly covered. A mirror in the room can be useful for the health care worker while donning PPE.

Preparing for doffing

The purpose of this step is to prepare for the removal of PPE. The doffing area should be separated into areas where early and later steps of doffing are conducted (e.g., separate chairs or ends of a bench). Before entering the PPE removal area, look for, clean, and disinfect (using an EPA-registered disinfectant wipe*) visible contamination on the PPE. As a final step before doffing, disinfect outer-gloved hands with either an EPA-registered disinfectant wipe* or ABHR, and allow to dry. Verify that the trained observer is available in the PPE removal area before entering and beginning the removal process. Some facilities, especially those using PAPRs, might find it helpful to have a designated assistant to help with doffing. An assistant who is only assisting in doffing should wear the same PPE as the trained observer. If the doffing assistant is entering the patient's room (e.g., as a clinician), the assistant should wear the same PPE as other personnel entering the patient's room. The observer should not touch the person who is doffing and should not serve as the doffing assistant or *buddy*. A mirror in the room can be useful for the health care worker while doffing PPE.

Section 9C. Doffing PPE, PAPR option

PPE should be doffed in the designated PPE removal area. Place all PPE waste in a leakproof infectious waste container.

1. *Engage trained observer*: The doffing process should be supervised by the trained observer, who reads aloud each step of the procedure and confirms visually that the PPE is removed properly. Before the health care worker doffs PPE, the trained observer should coach and

remind the health care worker to avoid reflexive actions that may put them at risk, such as touching their face. Post this instruction and repeat it verbally during doffing.

2. *Inspect*: Inspect the PPE to assess for visible contamination, cuts, or tears before starting to remove. If any PPE is visibly contaminated, then clean and disinfect using an EPA-registered disinfectant wipe.

3. *Disinfect outer gloves*: Disinfect outer-gloved hands with either an EPA-registered disinfectant wipe* or ABHR, and allow to dry.

4. *Remove apron (if used)*: Remove (e.g., by breaking or untying neck strap and releasing waist ties) and roll the apron away from you, containing the soiled outer surface as you roll; discard apron taking care to avoid contaminating gloves or other surfaces.

5. *Inspect*: After removing the apron, inspect the PPE ensemble for visible contamination or cuts or tears. If visibly contaminated, then clean and disinfect affected areas using an EPA-registered disinfectant wipe.*

6. *Disinfect and remove outer gloves*: Disinfect outer-gloved hands with either an EPA-registered disinfectant wipe* or ABHR. Remove and discard outer gloves, taking care not to contaminate inner glove during removal process.

7. *Inspect and disinfect inner gloves*: Inspect the inner gloves' outer surfaces for visible contamination, cuts, or tears. If an inner glove is visibly soiled, then disinfect the glove with either an EPA-registered disinfectant wipe* or ABHR, remove the inner gloves, perform hand hygiene with ABHR on bare hands, and don a new pair of gloves. If no visible contamination is identified on the inner gloves, then disinfect the inner gloves with either an EPA-registered disinfectant wipe* or ABHR. If a cut or tear is detected on an inner glove, immediately review occupational exposure risk per hospital protocol.

8. *Remove respirator with external belt-mounted blower*: Remove the headpiece. The health care worker may need help for removing the headpiece while still connected to the belt-mounted blower and filter unit. (Note: If a PAPR with a self-contained blower in the helmet is used, wait until step 14 to remove components.)

   a. Remove the belt-mounted blower unit and place all reusable PAPR components in an area or container designated for the collection of PAPR components for disinfection.

   b. Disinfect inner gloves with either an EPA-registered disinfectant wipe* or ABHR.

9. *Remove gown* or *coverall*: Remove and discard.

   a. Depending on gown design and location of fasteners, the health care worker can either untie fasteners, have the doffing assistant or *buddy* unfasten the gown, or gently break fasteners. Avoid contact of scrubs or disposable garments with outer surface of gown during removal. Pull gown away from body, rolling inside out and touching only the inside of the gown.

   b. To remove coverall, tilt head back and reach zipper or fasteners. Use a mirror to avoid contaminating skin or inner garments. Unzip or unfasten coverall completely before rolling down and turning inside out. Avoid contact of scrubs with outer surface of coverall during removal, touching only the inside of the coverall.

10. *Disinfect inner gloves*: Disinfect inner gloves with either an EPA-registered disinfectant wipe* or ABHR.

11. *Remove boot covers*: Sitting on a new clean surface (e.g., second clean chair and clean side of a bench) pull off boot covers, taking care not to contaminate pant legs.

12. *Disinfect washable shoes*: Use an EPA-registered disinfectant wipe* to wipe down every external surface of the washable shoes.

13. *Disinfect inner gloves*: Disinfect inner gloves with either an EPA-registered disinfectant wipe* or ABHR.

14. *Remove respirator (if not already removed)*: If a PAPR with a self-contained blower in the helmet is used, remove all remaining components here.

   a. Remove and discard disposable hood.

   b. Disinfect inner gloves with either an EPA-registered disinfectant wipe* or ABHR.

   c. Remove helmet and the belt and battery unit. The health care worker may need help removing the PAPR.

   d. Place all reusable PAPR components in an area or container designated to collect PAPR components for disinfection.

15. *Disinfect and remove inner gloves*: Disinfect inner-gloved hands with either an EPA-registered disinfectant wipe* or ABHR. Remove and discard gloves, taking care not to contaminate bare hands during removal process.

16. *Perform hand hygiene*: Perform hand hygiene with ABHR.

17. *Inspect*: Both the trained observer and the health care worker perform a final inspection of the health care worker for contamination of surgical scrubs or disposable garments. If contamination is identified, the garments should be carefully removed and the wearer should shower immediately. The trained observer should immediately

inform the infection preventionist or occupational safety and health coordinator or their designee for appropriate occupational health follow-up.

18. *Scrubs*: Health care worker can leave the PPE removal area wearing dedicated washable footwear and surgical scrubs or disposable garments, proceeding directly to showering area where these are removed.

19. *Protocol evaluation/medical assessment*: Either the infection preventionist or occupational safety and health coordinator or their designee should meet with each health care worker on a regular basis to review the patient care activities performed, identify any concerns about care protocols, and record the health care worker's level of fatigue.

Section 9D. Doffing PPE, N95 respirator option
PPE should be doffed in the designated PPE removal area. Place all PPE waste in a leakproof infectious waste container.

1. *Engage trained observer*: The doffing process should be supervised by the trained observer, who reads aloud each step of the procedure and confirms visually that the PPE has been removed properly. Before doffing PPE, the trained observer must remind health care workers to avoid reflexive actions that may put them at risk, such as touching their face. Post this instruction and repeat it verbally during doffing.

2. *Inspect*: Inspect the PPE to assess for visible contamination, cuts, or tears before starting to remove. If any PPE is visibly contaminated, then disinfect using an EPA-registered disinfectant wipe.*

3. *Disinfect outer gloves*: Disinfect outer-gloved hands with either an EPA-registered disinfectant wipe* or ABHR.

4. *Remove apron (if used)*: Remove (e.g., by breaking or untying neck strap and releasing waist ties) and roll the apron away from you, containing the soiled outer surface as you roll; discard apron taking care to avoid contaminating gloves or other surfaces.

5. *Inspect*: After removing the apron, inspect the PPE ensemble for visible contamination or cuts or tears. If visibly contaminated, then clean and disinfect any affected areas by using an EPA-registered disinfectant wipe.*

6. *Disinfect and remove outer gloves*: Disinfect outer-gloved hands with either an EPA-registered disinfectant wipe* or ABHR. Remove and discard outer gloves, taking care not to contaminate inner gloves during removal process.

7. *Inspect and disinfect inner gloves*: Inspect the inner gloves' outer surfaces for visible contamination, cuts, or tears. If an inner glove is visibly soiled, then disinfect the glove with either an EPA-registered disinfectant wipe* or ABHR, remove the inner gloves, perform hand hygiene with ABHR on bare hands, and don a new pair of gloves. If no visible contamination is identified on the inner gloves, then disinfect the inner-gloved hands with either an EPA-registered disinfectant wipe* or ABHR. If a cut or tear is detected on an inner glove, immediately review occupational exposure risk per hospital protocol.

8. *Remove face shield*: Remove the full-face shield by tilting the head slightly forward, grasping the rear strap and pulling it gently over the head and allowing the face shield to fall forward, then discard. Care must be taken not to touch the face when removing the face shield. Avoid touching the front surface of the face shield.

9. *Disinfect inner gloves*: Disinfect inner gloves with either an EPA-registered disinfectant wipe* or ABHR.

10. *Remove surgical hood*: Unfasten (if applicable) surgical hood, gently remove, and discard. The doffing assistant or "buddy" can assist with unfastening hood.

11. *Disinfect inner gloves*: Disinfect inner gloves with either an EPA-registered disinfectant wipe* or ABHR.

12. *Remove gown or coverall*: Remove and discard.
    a. Depending on gown design and location of fasteners, the health care worker can untie fasteners, have the doffing assistant or *buddy* unfasten the gown, or gently break fasteners. Avoid contact of scrubs or disposable garments with outer surface of gown during removal. Pull gown away from body, rolling inside out and touching only the inside of the gown.
    b. To remove coverall, tilt head back to reach zipper or fasteners. Unzip or unfasten coverall completely before rolling down and turning inside out. Avoid contact of scrubs with outer surface of coverall during removal, touching only the inside of the coverall.

13. *Disinfect inner gloves*: Disinfect inner gloves with either an EPA-registered disinfectant wipe* or ABHR.

14. *Remove boot covers*: Sitting on a clean surface (e.g., second clean chair or clean side of a bench) pull off boot covers, taking care not to contaminate scrubs pant legs.

15. *Disinfect and change inner gloves*: Disinfect inner gloves with either an EPA-registered disinfectant wipe* or ABHR.
    a. Remove and discard gloves taking care not to contaminate bare hands during removal process.

b. Perform hand hygiene with ABHR.

c. Don a new pair of inner gloves.

16. *Remove N95 respirator*: Remove the N95 respirator by tilting the head slightly forward, grasping first the bottom tie or elastic strap, then the top tie or elastic strap, and remove without touching the front of the N95 respirator. Discard N95 respirator.

17. *Disinfect inner gloves*: Disinfect inner gloves with either an EPA-registered disinfectant wipe* or ABHR.

18. *Disinfect washable shoes*: Use an EPA-registered disinfectant wipe* to wipe down every external surface of the washable shoes.

19. *Disinfect and remove inner gloves*: Disinfect inner-gloved hands with either an EPA-registered disinfectant wipe* or ABHR. Remove and discard gloves taking care not to contaminate bare hands during removal process.

20. *Perform hand hygiene*: Perform hand hygiene with ABHR.

21. *Inspect*: Both the trained observer and the health care worker perform a final inspection of health care worker for contamination of the surgical scrubs or disposable garments. If contamination is identified, the garments should be carefully removed and the wearer should shower immediately. The trained observer should immediately inform infection preventionist or occupational safety and health coordinator or their designee.

22. *Scrubs*: Health care worker can leave PPE removal area wearing dedicated washable footwear and surgical scrubs or disposable garments, proceeding directly to the showering area where these are removed.

23. *Protocol evaluation/medical assessment*: Either the infection preventionist or occupational health safety and health coordinator or their designee should meet with the health care worker on a regular basis to review the patient care activities performed, identify any concerns about care protocols, and record health care worker's level of fatigue.

REFERENCE

Centers for Disease Control and Prevention (CDC). Guidance on personal protective equipment (PPE) to be used by healthcare workers during management of patients with confirmed Ebola or persons under investigation (PUIs) for Ebola who are clinically unstable or have bleeding, vomiting, or diarrhea in U.S. hospitals, including procedures for donning and doffing. August 27, 2015. https://www.cdc.gov/vhf/ebola/healthcare-us/ppe/guidance.html.

CENTERS OF DISEASE CONTROL AND PREVENTION GUIDELINES ON EBOLA RISK FACTORS

What follows are the CDC's guidelines on Ebola risk factors, as of May 28, 2015.

Guidelines for Epidemiologic Risk Factors to Consider When Evaluating a Person for Exposure to Ebola Virus

The following epidemiologic risk factors should be considered when evaluating a person for Ebola virus disease (EVD), classifying contacts, or considering public health actions such as monitoring and movement restrictions based on exposure.

1. *High risk* includes any of the following:
   In any country
   a. Percutaneous (e.g., needle stick) or mucous membrane exposure to blood or body fluids (including but not limited to feces, saliva, sweat, urine, vomit, and semen*) from a person with Ebola who has symptoms.
   b. Direct contact with a person with Ebola who has symptoms, or the person's body fluids, *while not wearing appropriate personal protective equipment (PPE)*.
   c. Laboratory processing of blood or body fluids from a person with Ebola who has symptoms *while not wearing appropriate PPE or without using standard biosafety precautions.*
   d. Providing direct care to a person showing symptoms of Ebola in a household setting.
   In countries with widespread transmission or cases in urban settings with uncertain control measures
   a. Direct contact with a dead body *while not wearing appropriate PPE.*

---

* Ebola virus can be detected in semen for months after recovery from the disease. Unprotected contact with the semen of a person who has recently recovered from Ebola may constitute a potential risk for exposure. The period of risk is not yet defined.

2. *Some risk* includes any of the following:
In any country
   a. Being in close contact* with a person with Ebola who has symptoms *while not wearing appropriate PPE* (e.g., in households, health care facilities, or community settings).
In countries with widespread transmission
   a. Direct contact with a person with Ebola who has symptoms, or the person's body fluids, *while wearing appropriate PPE*.
   b. Being in the patient-care area of an Ebola treatment unit.
   c. Providing any direct patient care in non-Ebola health care settings.
3. *Low (but not zero) risk* includes any of the following:
In any country
   a. Brief direct contact (such as shaking hands) with a person in the early stages of Ebola, *while not wearing appropriate PPE*. Early signs can include fever, fatigue, or headache.
   b. Brief proximity with a person with Ebola who has symptoms (such as being in the same room, but not in close contact) *while not wearing appropriate PPE*.
   c. Laboratory processing of blood or body fluids from a person with Ebola who has symptoms *while wearing appropriate PPE and using standard biosafety precautions*.
   d. Traveling on an airplane with a person with Ebola who has symptoms and having had no identified *some* or *high* risk exposures.
In countries with widespread transmission, cases in urban settings with uncertain control measures, or former widespread transmission and current established control measures are as follows:
   a. Having been in one of these countries and having had no known exposures.
In any country other than those with widespread transmission
   a. Direct contact with a person with Ebola who has symptoms, or the person's body fluids, while wearing appropriate PPE.
   b. Being in the patient-care area of an Ebola treatment unit.
4. *Nonidentifiable risk* includes any of the following:
   a. Laboratory processing of Ebola-containing specimens in a Biosafety Level 4 facility.
   b. Any contact with a person who is not showing symptoms of Ebola, even if the person had potential exposure to Ebola virus.

---

* Close contact is defined as being within approximately three feet (one meter) of a person with Ebola while the person was symptomatic for a prolonged period of time *while not using appropriate PPE*.

   c. Contact with a person with Ebola before the person developed symptoms.

   d. Any potential exposure to Ebola virus that occurred *more than 21 days previously.*

   e. Having been in a country with Ebola cases, but *without* widespread transmission, cases in urban settings with uncertain control measures, or former widespread transmission and now established control measures, and not having had any other exposures.

   f. Having stayed on or very close to an airplane or ship (e.g., to inspect the outside of the ship or plane or to load or unload supplies) during the entire time that the airplane or ship was in a country with widespread transmission or a country with cases in urban settings with uncertain control measures, *and* having had no direct contact with anyone from the community.

   g. Having had laboratory-confirmed Ebola and subsequently been determined by public health authorities to no longer be infectious (i.e., Ebola survivors).

REFERENCE

Centers for Disease Control and Prevention (CDC). Guidelines for epidemiologic risk factors to consider when evaluating a person for exposure to Ebola virus. August 28, 2015. https://scholar.google.com/scholar?q=Guidelines+for+epidemiologic+risk+factors+to+consider+when+evaluating+a+person+for+exposure+to+Ebola+virus.&btnG=&hl=en&as_sdt=0%2C33.

EBOLA TIMELINE, 1976–2016

The Ebola outbreak of West Africa that began in December 2013, rapidly reached historic proportions in terms of the number of cases and the number of countries involved (Briand et al. 2014). What follows demonstrates the rapidity of spread of the epidemic and the efforts aimed at containment from local, national, and global health authorities, which were ultimately successful. As can be seen from the summary of 24 prior outbreaks, Ebola had not been diagnosed in this region of Africa before and had never been seen in so many countries simultaneously. The countries involved witnessed spread of infection to their most densely populated cities. Efforts at containment were initially frustrated by a lack of medical facilities and personnel as well as by fears that medical centers represented a means of transmission of the infection. A large number of medical workers contracted infection and many died. This was due to a variety of factors including the lack of availability of effective personal protective equipment and the difficulty of imposing adequate infection control procedures. As shown in Map A.1, the epidemic spread quickly and involved an unprecedented number of victims. As international efforts, particularly by Doctors Without Borders (MSF) and, subsequently the U.S. Centers for Disease Control and Prevention (CDC) and the World Health Organization (WHO), accelerated the epidemic and came under control. The final totals of cases and deaths, although devastating, were much less than the numbers projected during the summer and fall of 2014. Nonetheless, the sudden appearance and rapid spread of a dangerous infection was not anticipated nor was the need for legions of health care workers to stop it.

The outbreak primarily involved three West African countries: Liberia, Sierra Leone, and Guinea. Other West African countries, Nigeria, Mali, and Senegal, saw small numbers of cases after the infection had been introduced by travelers from the original three countries. The United States received patients with Ebola evacuated or transported from West Africa, but also saw local transmission to two nurses in Dallas who had been exposed to a patient from West Africa. In addition, one U.S. physician was diagnosed with Ebola only after returning from West Africa.

The spread of Ebola beyond an initial small region as seen in this outbreak was unique. Addressing this fact as well as establishing containment and, finally, bringing the epidemic under control in the absence of an effective vaccine ultimately represented an enormous international effort.

How the Epidemic Grew

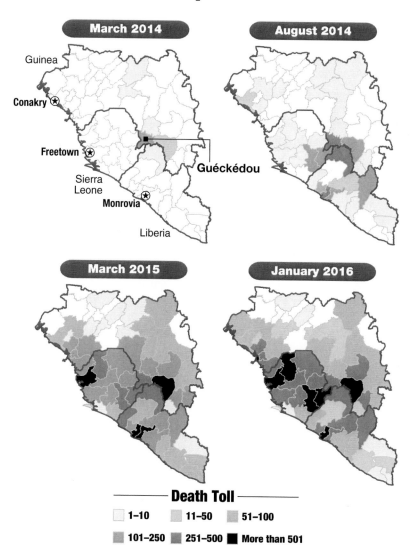

MAP A.1   How the epidemic spread. (Data courtesy of WHO, Geneva, Switzerland. Map by Rod Eyer.)

This timeline includes information about outbreaks of EVD since the virus was first recognized as a human pathogen in 1976, concluding with the World Health Organization's declaration in January 2016 that the epidemic had ended. Scattered cases occurred later that spring, but the outbreak did not undergo a resurgence.

1976–2014 Cases in East and Southern Africa

This section of the timeline draws largely on the following references: Peters and LeDuc 1999; Streether 1999; Colebunders and Borchert 2000; Leroy et al. 2011; MacNeil and Rollin 2012; Shoemaker et al. 2012; Weyer et al. 2015; Bell et al. 2016.

Prior to the 2014–2016 outbreak, Ebola had been seen in East and South Africa in 24 small to moderate outbreaks beginning in 1976 (Breman and Johnson 2014). Ebola virus disease (EVD) was recognized as a highly lethal but rare infection thought to be transmitted by an unknown animal vector to humans followed by brief periods of human-to-human transmission. It was regarded as a local threat to certain regions of Africa, a disease that had affected the great ape populations in these regions and that posed an international threat through imported infections and, potentially, bioterrorism (Feldmann and Geisbert 2011). The previous outbreaks had occurred in East and South Africa, regions not adjacent to the countries of West Africa that experienced the 2014–2016 outbreaks.

24 outbreaks in East and Southern Africa (CDC 2016b)
Total cases: 2411 (confirmed and suspected)
Total deaths: 1595 (66% mortality)

Outbreaks in seven countries of East and South Africa:

- Seven in Democratic Republic of Congo (1976, 1977, 1995, 2007, 2008–2009, 2012, 2014)
- Five in Uganda (2000, 2007, 2011, 2012, 2012)
- Four in Gabon (1994, 1996, 1996, 2001)
- Three in Republic of the Congo 2002, 2003, 2003 (two separate outbreaks)
- Three in South Sudan (1976, 1979, 2004)
- One in Cote d'Ivoire (1994)
- One in South Africa (1996)

2014–2016 in West Africa

What follows is a timeline of the 2014–2016 outbreaks in West Africa (CDC 2016b). The exact origin of the virus in each country is not known, but genetic sequencing of viral isolates suggests that the outbreak originated in Guinea and spread to Sierra Leone in the spring of 2014 (Carroll et al. 2015). The growth in cases in the three primary countries—Guinea, Liberia, and Sierra Leone, is shown in Figure A.1.

Total cases: 28,652 (confirmed and suspected)
Total deaths: 11,325 (39.5% mortality)

Cases in six countries of West Africa:

- 3814 in Guinea
- 10,678 in Liberia
- 14,124 in Sierra Leone
- 8 in Mali
- 20 in Nigeria
- 1 in Senegal

Cases initially diagnosed in three countries of Europe:

- One in Spain
- One in Italy
- One in the United Kingdom

Cases initially diagnosed in one country of North America:

- Four in the United States

Events of 2014–2016

The events of 2014–2016 involved 10 countries. What follows are key literature references to the Ebola outbreak in the countries that saw transmission, followed by a timeline of key events in the global outbreak.

Guinea (Baize et al. 2014; Bah et al. 2015; Faye et al. 2015; Hersey et al. 2015; Thiam et al. 2015; Victory 2015; Kpanake et al. 2016; Lindblade et al. 2016; Rico at al. 2016)

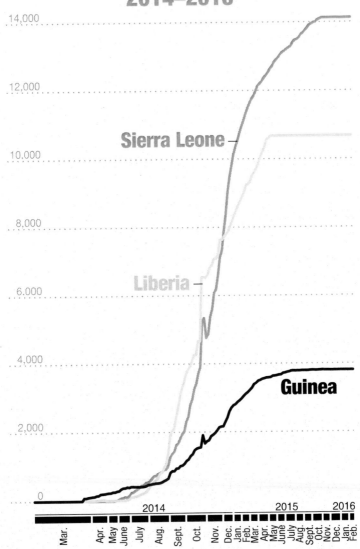

FIGURE A.1 This graph shows the size of the outbreak in the most-affected nations.

Liberia (Massaquoi et al. 2014; Matanock et al. 2014; Nyenswah et al. 2014, 2016; Christie et al. 2015; Yamin et al. 2015; Atkins et al. 2016; Cooper et al. 2016)

Sierra Leone (Kilmarx et al. 2014; Schieffelin et al. 2014; Dietz et al. 2015; Lado et al. 2015; Lu et al. 2015; Qin et al. 2015; Yan et al. 2015; Curran et al. 2016; Mattia et al. 2016)

Nigeria (Shuaib et al. 2014; Althaus et al. 2015; Grigg et al. 2015; Vaz et al. 2016)

Mali (Hoenen et al. 2015)

Senegal (Mirkovic et al. 2014)

United States (Benowitz et al. 2014; Chevalier et al. 2014; Regan et al. 2015; Smith et al. 2015; Yacisin et al. 2015; Uyeki et al. 2016)

Spain (Lopaz et al. 2015)

Key Events in the Outbreak

December 28, 2013: A two-year-old child, his mother, sister, and grandmother died of hemorrhagic fever. Mourners at the funeral are thought to have contracted infection and, subsequently, carried the virus to nearby villages (Baize et al. 2014). It is thought that the infection spread for several months, potentially misdiagnosed as other diseases endemic to West Africa, before it was recognized as Ebola.

March 23, 2014: First Ebola cases of the West Africa outbreak are reported in southeastern Guinea in areas bordering Sierra Leone and Liberia.

March 28, 2014: At the end of March, the WHO reported that there had been 112 cases and 70 deaths due to Ebola in West Africa, including two confirmed cases in Liberia (WHO 2014a).

April 30, 2014: The Ministry of Health of Guinea reported that 26 of 221 cases were in health care workers.

PHOTO A.2   Ebola treatment rooms in Guinea in 2014. The brown containers contain chlorinated water for hand washing. They are marked *ALIMA*, for the Alliance for International Medical Action, a nongovernmental organization based in Senegal and France that participated in relief efforts. (Courtesy of Dr. Heidi Soeters, CDC, Atlanta, GA.)

May 12, 2014: Cases are reported in Conakry, the capital of Guinea.

May 26, 2014: First cases are reported in Sierra Leone (Dahl et al. 2016).

June 11, 2014: Sierra Leone closes its borders with Guinea.

June 17, 2014: Liberia reports cases in its capital, Monrovia.

July 25, 2014: Liberian government employee dies in Nigeria of Ebola.

July 27, 2014: Liberia shuts down most border crossings (BBC 2014).

July 29, 2014: Dr. Sheik Umar Khan, who had played a key role in combating the spread of Ebola infection in Sierra Leone, dies.

July 30, 2014: The United States announces it will withdraw all Peace Corps volunteers from Liberia, Guinea, and Sierra Leone because of Ebola risks (U.S. Peace Corps 2014).

August 2, 2014: An American missionary aid worker, Dr. Kent Brantly, infected with Ebola in Liberia, is evacuated to Atlanta, Georgia, and treated at Emory University Hospital.

August 5, 2014: A second missionary aid worker, Nancy Writebol, is evacuated to Emory University Hospital, Atlanta, Georgia.

August 7, 2014: World Health Organization (WHO) declares the outbreak a "Public Health Emergency of International Concern." (PHEIC) (WHO 2014b)

August 12, 2014: Death toll exceeds 1000.

August 21, 2014: The first two medically evacuated cases, Brantly and Writebol, who were treated at Emory University Hospital with the experimental therapy ZMapp, are released from the hospital.

August 21, 2014: Ivory Coast closes its borders with Guinea and Liberia.

August 22, 2014: A physician in Port Harcourt, Nigeria, Ikechukwu Enemuo dies of Ebola. This is the second death from Ebola in Nigeria.

August 24, 2014: A British nurse serving as a volunteer in Sierra Leone, Will Pooley, is flown back to the United Kingdom after contracting Ebola. He recovers and is released from the hospital on September 3, 2014.

August 27, 2014: A Senegalese epidemiologist working for the WHO from Sierra Leone is transferred to an isolation ward in Hamburg, Germany, with Ebola.

August 29, 2014: Riots break out in Guinea following rumors that Ebola is being intentionally spread by health care workers (Camara 2014).

Senegal reports a case

September 5, 2014: A Massachusetts physician, Richard Sacra, who had been working in Liberia performing caesarean sections on women with Ebola was brought to the Nebraska Medical Center in Omaha with Ebola. He received a transfusion from the patient, Dr. Kent Brantly, who had recovered at Emory University Hospital.

WHO estimates 4000 cases and 2100 deaths in the current outbreak

September 9, 2014: A physician who had been working in Sierra Leone for WHO is brought to Emory University Hospital in Atlanta for treatment.

September 16, 2014: President Barack Obama announces that the United States will send 3000 troops to West Africa to build treatment centers and a coordination center.

September 18, 2014: The United Nations General Assembly and Security Council approve the Nations Mission for Ebola Emergency Response (UNMEER).

September 19, 2014: A French nurse volunteer with Doctors Without Borders with EVD is evacuated from Liberia to France. She recovers and is released on October 4.

September 22, 2014: A new 150-bed treatment center is opened in Monrovia.

September 26, 2014: The WHO estimates 6574 cases and 3091 deaths.

September 29, 2014: Thomas Eric Duncan, a Liberian, flies from Monrovia to Dallas via Brussels and dies of Ebola in a Dallas hospital just over one week later, becoming the first person diagnosed with Ebola in the Western hemisphere. Two Dallas hospital nurses become infected, becoming the first persons to contract Ebola in the Western hemisphere (CDC 2014).

October 7, 2014: A Spanish nurse in Madrid, Teresa Romero Ramos, contracts Ebola from a Spanish priest who had been flown to Spain for treatment (Lopaz et al. 2015).

October 17, 2014: The Ebola outbreak in Senegal is declared over after no new cases were reported in 42 days (WHO 2014c).

October 23, 2014: Physician Craig Spencer, who had recently returned from Guinea where he had been treating patients with EVD with Doctors Without Borders, is placed in an isolation unit at Bellevue Hospital in New York City after experiencing symptoms of EVD. He subsequently tests positive for Ebola virus (Spencer 2015).

October 25, 2014: WHO announces that the EVD outbreak has surpassed 10,000 cases, with 4922 deaths (Azikiwe 2014).

November 15, 2014: A surgeon, Dr. Martin Salia, is evacuated from Sierra Leone to the Nebraska Medical Center. He dies on November 17, 2014 of EVD.

January 3, 2015: Confirmed deaths reach 8000.

January 8, 2015: Confirmed cases reach 21,000.

January 18, 2015: Mali is declared Ebola-free (WHO 2015a).

April 1, 2015: Total cases exceed 25,000. Deaths exceed 10,000 (Schnirring 2015).

December 29, 2015: Guinea is declared Ebola-free, meaning transmission of Ebola has ended (WHO 2015b).

January 14, 2016: Liberia is declared Ebola-free (WHO 2016a).

West Africa is declared Ebola-free (WHO 2016a)

PHOTO A.3   Schools in Guinea, similar to this one in the capital, Conakry, were closed for eight months because of the Ebola epidemic. When they reopened in January 2015, attendance was sparse. (Courtesy of Alex Alvarez, CDC/CDC Connects.)

March and April 2016: About a dozen additional cases of Ebola occur in Guinea and Liberia.

June 1, 2016: Guinea is again declared Ebola-free (WHO 2016b).

June 9, 2016: Liberia again is declared Ebola-free (WHO 2016c).

REFERENCES

Althaus CL, Low N, Musa EO, Shuaib F, Gsteiger S. Ebola virus disease outbreak in Nigeria: Transmission dynamics and rapid control. *Epidemics* 2015;11:80–84.

Atkins KIE, Pandey A, Wenzel NS et al. Retrospective analysis of the 2014–2015 Ebola epidemic in Liberia. *Am J Trop Med Hyg* 2016;94(4):833–839.

Azikiwe A. WHO announces that the EVD outbreak has surpassed 10,000 cases, with 4,922 deaths. *Global Research*, October 27, 2014. http://www.globalresearch. ca/ebola-cases-surpass-10000-in-west-africa-while-united-states-politicians-impose-quarantines/5410227.

Bah EI, Fletcher T, Jacob ST et al. Clinical presentation of patients with Ebola virus disease in Conakry, Guinea. *N Engl J Med* 2015;372(1):40–47.

Baize S, Pannetier D, Oestereich L et al. Emergence of Zaire Ebola virus disease in Guinea. *N Engl J Med* 2014;371(15):1418–1425.

BBC. Ebola outbreak: Liberia shuts most border points. July 28, 2014. http://www.bbc.com/news/world-africa-28522824.

Bell BP, Damon IK, Jernigan DB et al. Overview, control strategies, and lessons learned in the CDC response to the 2014–16 Ebola epidemic. *MMWR* 2016;65(3):4–11.

Benowitz I, Ackelsberg J, Balter SE et al. Surveillance and preparedness for Ebola virus disease—New York City, 2014. *MMWR* 2014;63(41):934–936.

Breman JG, Johnson KM. Ebola then and now. *N Engl J Med* 2014;371(18):1663–1666.

Briand S, Bertherat E, Cox P et al. The international Ebola emergency. *N Engl J Med* 2014;371(13):1180–1182.

Camara O. Ebola riots in Guinea leave seven dead, premier says. September 18, 2014. https://www.bloomberg.com/news/articles/2014-09-18/ebola-riots-in-guinea-leave-seven-dead-premier-says.

Carroll MW, Matthews DA, Hiscox JA et al. Temporal and spatial analysis of the 2014–2015 Ebola virus outbreak in West Africa. *Nature* 2015;524(7563):97–101.

Centers for Disease Control and Prevention (CDC). Cases of Ebola diagnosed in the United States. December 16, 2014. https://www.cdc.gov/vhf/ebola/outbreaks/2014-west-africa/united-states-imported-case.html.

Centers for Disease Control and Prevention (CDC). 2014 Ebola outbreak in West Africa case counts. April 13, 2016a. www.cdc.gov/vhf/ebola/outbreaks/2014-west-africa/case-counts/html.

Centers for Disease Control and Prevention (CDC). Outbreaks chronology: Ebola virus disease. April 14, 2016b. https://www.cdc.gov/vhf/ebola/outbreaks/history/chronology.html.

Chevalier MS, Chung W, Smith J et al. Ebola virus cluster in the United States—Dallas County, Texas, 2014. *MMWR* 2014;63(46):1087–1088.

Christie A, Vertefeuille J, Olsen SG et al. Controlling the last know cluster of Ebola virus disease—Liberia, January–February 2015. *MMWR* 2015;64(18):500–504.

Colebunders R, Borchert M. Ebola hemorrhagic fever—A review. *J Infect* 2000; 40(1):16–20.

Cooper C, Fisher D, Gupta N, MaCauley R, Pessoa-Silva CL. Infection prevention and control of the Ebola outbreak in Liberia, 2014–2015: Key challenges and successes. *BMC Med* 2016;24:2.

Curran KG, Gibson JJ, Marke D et al. Cluster of Ebola virus disease linked to a single funeral—Moyamba District, Sierra Leone, 2014. *MMWR* 2016;65(8):202–205.

Dahl BA, Kinzer MH, Raghunathan PL et al. CDC's response to the 2014–2016 Ebola epidemic—Guinea, Liberia, and Sierra Leone. *MMWR* 2016;65(3):12–20.

Dietz PM, Jambai A, Paweska JT, Yoti Z, Ksiazek TG. Epidemiology and risk factors for Ebola virus disease in Sierra Leone—May 23, 2014 to January 31, 2015. *Clin Infect Dis* 2015;61(11):1648–1654.

Faye O, Boalle PY, Heleze E et al. Chains of transmission and control of Ebola virus disease in Conakry, Guniea in 2014: An observational study. *Lancet Infect Dis* 2015;15(3):320–326.

Feldmann H, Geisbert TW. Ebola haemorrhagic fever. *Lancet* 2011;377(9768): 849–862.

Grigg C, Waziri NE, Olayinka AT, Vertefeuille JF. Use of group quarantine in Ebola control—Nigeria, 2014. *MMWR* 2015;64(5):124.

Hersey S, Martel LD, Jambi A et al. Ebola virus disease—Sierra Leone and Guinea, August, 2015. *MMWR* 2015;64(35):981–984.

Hoenen T, Safronetz D, Groseth A et al. Virology. Mutation rate and genotype variation of Ebola virus from Mali case sequences. *Science* 2015;348(6230):117–119.

Kilmarx PH, Clarke KR, Dietz PM et al. Ebola virus disease in health care workers— Sierra Leone, 2014. *MMWR* 2014;63(49):1168–1171.

Kpanake L, Gossou K, Sorum PC, Mullet E. Misconceptions about Ebola virus disease among lay people in Guinea: Lessons for community education. *J Public Health Policy* 2016;37(2):160–172.

Lado M, Walker NJ, Baker P et al. Clinical features of patients isolated for suspected Ebola virus disease at Connaught Hospital, Freetown, Sierra Leone: A retrospective cohort study. *Lancet Infect Dis* 2015;15(9):1024–1033.

Leroy EM, Bonzalez JP, Baize S. Ebola and Marburg haemorrhagic fever viruses: Major scientific advances, but a relatively minor public health threat for Africa. *Clin Microbiol Infect* 2011;17(7):964–976.

Lindblade KA, Nyenswah T, Keita S et al. Secondary infections with Ebola virus in rural communities, Liberia and Guinea, 2014–2015. *Emerg Infect Dis* 2016; 22(9):1653–1655.

Lopaz MA, Amela C, Ordobas M et al. First secondary case of Ebola outside Africa: Epidemiological characteristics and contact monitoring, Spain, September to November 2014. *Euro Surveill* 2015;20(1).

Lu HJ, Qian J, Kargbo D et al. Ebola virus outbreak investigation, Sierra Leone, September 28–November 11, 2014. *Emerg Infect Dis* 2015;21(11):1921–1927.

MacNeil A, Rollin PE. Ebola and Marburg hemorrhagic fevers: Neglected tropical diseases? *PLoS Negl Trop Dis* 2012;6(6):e1546.

Massaquoi M, Nagbe T, Bawo L et al. Ebola epidemic—Liberia, March–October 2014. *MMWR* 2014;63(46):1082–1086.

Matanock A, Arwady MA, Ayscue P et al. Ebola virus disease cases among health care workers not working in Ebola treatment units—Liberia, June–August 2014. *MMWR* 2014;63(46):1077–1081.

Mattia JG, Vandy MJ, Chang JC et al. Early clinical sequelae of Ebola virus disease in Sierra Leone: A cross-sectional study. *Lancet Infect Dis* 2016;16(3): 331–338.

Mirkovic K, Thwing J, Diack PA. Importation and containment of Ebola virus disease— Senegal, August–September 2014. *MMWR* 2014;63(39):873–874.

Nyenswah T, Fahnbulleh M, Massaquoi M et al. Ebola epidemic—Liberia, March– October 2014. *MMWR* 2014;63:1–5.

Nyenswah TG, Kateh F, Bawo L et al. Ebola and its control in Liberia, 2014–2015. *Emerg Infect Dis* 2016;22(2):169–177.

Peters CJ, LeDuc JW. An introduction to Ebola: The virus and the disease. *J Infect Dis* 1999;179(Suppl 1):ix–xvi.

Qin E, Bi J, Zhao M et al. Clinical features of patients with Ebola virus disease in Sierra Leone. *Clin Infect Dis* 2015;61(4):491–495.

Regan JJ, Jungerman R, Montiel SH et al. Public health response to commercial airline travel of a person with Ebola virus infection—United States, 2014. *MMWR* 2015;64(3):63–66.

Rico A, Brody D, Coronado F et al. Epidemiology of Ebola virus disease in Conakry and surrounding prefectures, Guinea, 2014–2015. *Emerg Infect Dis* 2016;22(2):178–183.

Schieffelin JS, Shaffer JG, Goba A et al. Clinical illness and outcomes in patients with Ebola in Sierra Leone. *N Engl J Med* 2014;371(22):2092–2100.

Schnirring L. Abstinence, other steps for Ebola nations as cases top 25,000. *CIDRAP News*, March 30, 2015.

Shoemaker T, MacNeil A, Balinandi S et al. Reemerging Sudan Ebola virus disease in Uganda, 2011. *Emerg Infect Dis* 2012;18(9):1480–1483.

Shuaib F, Gunnala R, Musa EO et al. Ebola virus disease outbreak—Nigeria, July–September 2014. *MMWR* 2014;63(39):867–872.

Smith CL, Hughes SM, Karwowski MP et al. Addressing needs of contacts of Ebola patients duringa an investigation of an Ebola cluster in the United States—Dallas, Texas, 2014. *MMWR* 2015;64(5):121–123.

Spencer C. Having and fighting Ebola—Public health lessons from a clinician turned patient. *N Engl J Med* 2015;372(12):1089–1091.

SteelFisher GK, Blendon RJ, Lasala-Blanco N. Ebola in the United States—Public reactions and implications. *N Engl J Med* 2015;373(9):789–791.

Streether LA. Ebola virus. *Br J Biomed Sci* 1999;56(4):280–284.

Thiam S, Delamou A, Camara S et al. Challenges in controlling the Ebola outbreak in two prefectures in Guinea: Why did communities continue to resist? *Pan Afr Med J* 2015;22(Suppl 1):22.

U.S. Peace Corps. Peace Corps removing volunteers in Liberia, Sierra Leone and Guinea. July 30, 2014. https://www.peacecorps.gov/news/library/peace-corps-removing-volunteers-in-liberia-sierra-leone-and-guinea/.

Uyeki TM, Mehta AK, Davey RT et al. Clinical management of Ebola virus disease in the United States and Europe. *N Engl J Med* 2016;374(7):636–646.

Vaz RG, Mkanda P, Banda R et al. The role of the polio program infrastructure in response to Ebola virus disease in Nigeria 2014. *J Infect Dis* 2016;213(Suppl 3): S140–S146.

Victory KR, Coronado F, Ifono SO, Soropogui T, Dahl BA. Ebola transmission linked to a single traditional funeral ceremony—Kissidoubou, Guinea, December 2014–January 2015. *MMWR* 2015;64(14):386–388.

Weyer J, Grobbelaar R, Blumberg L. Ebola virus disease: History, epidemiology and outbreaks. *Curr Infect Dis Rep* 2015;17(5):480.

World Health Organization (WHO). Ebola virus disease in Guinea—Update. March 27, 2014a. http://www.who.int/csr/don/2014_03_27_ebola/en/.

World Health Organization (WHO). Statement on the 1st meeting of the IHR Emergency Committee on the 2014 Ebola outbreak in West Africa. August 8, 2014b. http://who.int/mediacentre/news/statements/2014/ebola-20140808/en/.

World Health Organization (WHO). The outbreak of Ebola virus disease in Senegal is over. October 17, 2014c. http://www.who.int/mediacentre/news/ebola/17-october-2014/en/.

World Health Organization (WHO). Government of Mali and WHO announce the end of the Ebola outbreak in Mali. January 18, 2015a. http://www.afro.who.int/en/media-centre/pressreleases/item/7293-government-of-mali-and-who-announce-the-end-of-the-ebola-outbreak-in-mali.html.

World Health Organization (WHO). End of Ebola transmission in Guinea. December 29, 2015b. http://www.afro.who.int/en/media-centre/pressreleases/item/8252-end-of-ebola-transmission-in-guinea.html.

World Health Organization (WHO). Latest Ebola outbreak over in Liberia; West Africa is at zero, but new flare-ups are likely to occur. January 14, 2016a. http://www.who.int/mediacentre/news/releases/2016/ebola-zero-liberia/en/.

World Health Organization (WHO). End of Ebola transmission in Guinea. June 1, 2016b. http://www.afro.who.int/en/media-centre/pressreleases/item/8676-end-of-ebola-transmission-in-guinea.html.

World Health Organization (WHO). End of Ebola transmission in Liberia. June 16, 2016c. http://www.afro.who.int/en/media-centre/pressreleases/item/8699-who-declares-the-end-of-the-most-recent-ebola-virus-disease-outbreak-in-liberia.html.

Yamin D, Gertler S, Ndeffo-Mbah ML et al. Effect of Ebola progression on transmission and control in Liberia. *Ann Intern Med* 2015;162(1):11–17.

Yacisin K, Balter S, Fine A et al. Ebola virus disease in a humanitarian aid worker—New York City, October 2014. *MMWR* 2015;64(12):321–323.

Yan T, Mu J, Qin E et al. Clinical characteristics of 154 patients suspected of having Ebola virus disease in the Ebola holding center of Jui Government Hospital in Sierra Leone during the 2014 Ebola outbreak. *Eur J Clin Microbiol Infect Dis* 2015;34(10):2089–2095.

EBOLA'S TOLL ON AFRICAN HEALTH WORKERS

During the West African epidemic, more than 800 health workers became infected with Ebola by March 2015, according to the World Health Organization. Of those whose outcome was known, two-thirds (418/635) died. It is possible, however, that the mortality rate may have been lower; for instance, outcomes may have been less likely to be recorded for workers who recovered. The WHO's definition of health workers for these tallies includes not only those delivering clinical case (e.g., doctors, nurses, nursing assistants, midwives, vaccinators, and medical students) but also pharmacy, laboratory, surveillance, mortuary, burial, and community health workers; ambulance drivers, janitors, guards, volunteers, and administrators. Those figures, like most figures in this book, include confirmed and probable cases. An additional 225 suspected cases in health workers were also reported (WHO 2015).

As presented in Table 1.5 of Chapter 1, by late May 2015, Guinea, Liberia, and Sierra Leone had lost a total of 240 doctors, nurses, and midwives to Ebola (Evans et al. 2015). Among the fatalities were key figures in the struggle to treat and contain the disease (Vogel 2014). Some of those lost were also involved in educating new physicians for their underresourced nations. The West African experience reinforces the crucial need for health care workers to have adequate training, full protective equipment, and access to other resources (Matanock et al. 2014).

It should be noted that some cases occurred in health workers who were not known to work in Ebola treatment or prevention activities; the source of infection often was unclear.

In addition to the physical toll it inflicted, Ebola damaged trust and bonds of social connection for many health workers. In one study done during the epidemic in Sierra Leone, workers at primary care units, not Ebola units, described feeling lonely, ostracized, afraid, sad, and deprived of trust and respect. The authors recommend that during an Ebola outbreak, health providers need psychological first aid. (McMahon et al. 2016). Probably the most intense loss came at Kenema, Sierra Leone. By September 2014, 26 staff members at the Kenema Government Hospital had died, including the nation's leading medical expert on viral hemorrhagic fevers, Dr. Sheik Humarr Khan, and the longtime chief nurse for its Lassa fever program, Mbalu Fonnie (AFP 2014; Hammer 2015). At that time, it was reported that of the 26 local nurses who had started working at the treatment center in May 2014, 19 had died of Ebola and five had been infected but survived

(Calkin 2014). By the end of the year, more than 40 nurses and doctors at Kenema had died (WHO 2015), and the facility had endured several strikes (Fofana and Giahyue 2014).

Here, in chronological order by date of death, are some of the nurses, physicians, and other health care workers who died of Ebola in 2014 during the worst days of the epidemic.

July 2014

Dr. Samuel Muhumuza Mutoro, 44, a Ugandan surgeon working at Redemption Hospital in New Kru Town, Liberia. He died in Monrovia, Liberia, on July 1.

Alex Moigboi, a nurse in the Lassa fever ward of Sierra Leone's Kenema Government Hospital, with more than 10 years experience caring for Lassa patients. He died there on July 19.

Alice Kovoma, a nurse in the Lassa ward of Sierra Leone's Kenema Government Hospital, where she had worked for six years.

Mbalu Fonnie, chief nurse of the Lassa fever ward and midwife at Sierra Leone's Kenema Government Hospital, where she had worked for 25 years. It is believed that nurses Fonnie, Moigboi, and Kovoma were all infected while caring for one pregnant patient. Nurse Fonnie died there on July 21. The next day, nurses at the hospital went on a strike.

Dr. Samuel Brisbane, 74, chief of the emergency room of the John F Kennedy Medical Center in Monrovia, Liberia, which he had helped to establish. Dr. Brisbane is said to have taught many of Liberia's doctors. He reportedly told the doctors treating him, "When we find ourselves in the middle of the sea and there are rough waves, we should not give up. We should fight on to the end" (Green 2014). He died in Monrovia on July 26.

Dr. Sheik Humarr Khan, 39, Sierra Leone's leading expert on clinical care of hemorrhagic fevers. Since 2005, he had been physician-in-charge of the Kenema Government Hospital's Lassa fever program. He died at the Doctors Without Borders (Médecins Sans Frontières) treatment center in Kailahun, Sierra Leone, on July 29. After his death, he was declared a national hero by the president of Sierra Leone.

Iye Gborie, a nurse in the Lassa fever ward at Kenema Government Hospital, Sierra Leone.

Mohamed Fullah, a lab technician who worked in the Lassa fever facility at Sierra Leon's Kenema Government Hospital and taught at the Eastern Polytechnic College in Sierra Leone.

August 2014

Brother Patrick Nshamdze, 52, director of the San José de Monrovia Hospital (St. Joseph's Catholic Hospital) in Monrovia, Liberia, and a member of Brothers Hospitallers of San Juan de Dios (Hospital Order of St. John of God). He was a native of Cameroon. He died at ELWA Hospital Monrovia on August 2.

Sister Rebecca Lansana, 42, a nurse at the Lassa fever ward at Sierra Leone's Kenema Government Hospital. She died during the first week of August.

Brother Miguel Pajares, 75, member of Brothers Hospitallers of San Juan de Dios (Hospital Order of St. John of God). He reportedly had cared for Brother Patrick Nshamdze. He was evacuated from Liberia to Madrid, Spain, where he died on August 12.

Dr. Modupeh John Horatio Cole, 56, a physician specialist in Connaught Hospital, Freetown, Sierra Leone. He died in Kailahun, Sierra Leone, at a treatment center run by Doctors Without Borders (Médecins Sans Frontières) on August 13.

Dr. Ameyo Stella Adadevoh, 57, an endocrinologist at First Consultants Medical Center in Lagos, Nigeria. Dr. Adadevoh is credited by many with having saved Nigeria from a catastrophic epidemic, by ordering an Ebola test for an ill traveler initially diagnosed with malaria, and insisting, despite the patient's strong objections, that he should stay in the hospital until the test results were known. The patient, Patrick Sawyer, died a few days later, on July 25. In the end, Dr. Adadevoh was one of only eight people who died of Ebola in Nigeria. She died in Lagos on August 19.

Dr. Ikechukwu Enemuo, a physician in Port Harcourt, Nigeria died on August 22.

Dr. Abraham Borbor, 54, an Internal medicine physician and deputy chief medical officer at John F Kennedy Medical Center in Monrovia, Liberia. He was responsible for the education of residents and interns at the medical center. He died in Monrovia on August 25.

Dr. Sahr Jimmy Rogers, a doctor at the Kenema Government Hospital. He died on August 30.

Hawa Rogers, a nurse at Kenema, was said to have cared for an orphaned baby whose mother had died of Ebola. She died on August 30.

September 2014

Sister Nancy Yoko, a nurse at the Ebola treatment center in Kenema. Her colleague Matron Josephine Sindesellu said that of the 26 local nurses that started working at the treatment center in May, Sister Yoko was the 19th to die from the virus (Calkin 2014).

Dr. Olivet Buck, 59, medical superintendent of Lumley Government Hospital, Freetown, Sierra Leone, where she was one of the only two doctors. Sierra Leone had asked funds from the WHO to transport Buck to Europe, saying the country could not afford to lose another doctor. WHO said it could not meet the request but instead would work to give Buck "the best care possible" in Sierra Leone (Barbash 2014). Dr. Buck died in Freetown on September 13.

Brother Manuel García Viejo, 69, Spanish physician and medical director at the San Juan de Dios in Lunsar, Sierra Leone. The hospital was closed after the death of nine health care workers. Brother Manuel was evacuated to Madrid, Spain, and died there on September 25.

October 2014

Dr. John Taban Dada, 55, gynecologist and surgeon at the JFK Medical Center in Monrovia, Liberia, formerly medical director of Monrovia's Redemption Hospital. He also taught postgraduate medical education. He died in Monrovia on October 9.

AbdelFadeel Mohammed Basheer, 56, medical laboratory technician at the UN Mission in Liberia. He was from Sudan. He was flown to Leipzig, Germany, for treatment and died there on October 14.

Dr. Thomas Scotland, a Liberian who received his medical degree in 2013. He was completing a medical internship and volunteering in the country's Ebola response. He died in Monrovia, Liberia on October 18.

November 2014

Dr. Godfrey George, 54, medical superintendent at Kambia Government Hospital in northwestern Sierra Leone and case manager for Kambia's local Ebola task force. He died in Freetown, Sierra Leone on November 3.

Dr. Martin Maada Salia, 44, a surgeon and chief medical officer of the Kissy United Methodist Hospital in Freetown, Sierra Leone. A native of Sierra Leone, Dr. Salia was a legal permanent resident of the United States and was the only African medically evacuated to the United States. He was flown to the Nebraska Medical Center's Ebola treatment unit, where he died two days later, on November 17.

Dr. Michael Moses Kargbo, 70, a surgeon in Sierra Leone who came out of retirement to join in the efforts against Ebola. He died at the Hastings Ebola Treatment Center, Freetown, Sierra Leone on November 18.

December 2014

Dr. Dauda Koroma died at the Hastings Ebola Treatment Center, Freetown, Sierra Leone on December 5.

Dr. Thomas Rogers, a surgeon at Connaught Hospital in Freetown, Sierra Leone. He died at the British-run Kerry Town Ebola treatment center on December 5.

Dr. Aiah Solomon Konoyeima, a doctor at a children's hospital in Freetown, Sierra Leone. He died at the Hastings Ebola Treatment Center on December 7.

Dr. Victor Willoughby, 67, Sierra Leone's most senior physician. In a tribute after his death he was called *the father of all doctors in Sierra Leone* (Sierra Leone Telegraph 2014). He died on December 18.

REFERENCES

The above list drew on the following sources, as well as accounts from numerous other news outlets:

AFP. Sierra Leone's 365 Ebola deaths traced back to one healer. *The Nation*, August 20, 2014. http://www.nation.co.ke/news/africa/Sierra-Leone-365-Ebola-deaths-traced-back-to-one-healer/1066-2424876-1427sxvz/index.html.

Barbash F. Sierra Leone loses fourth doctor to Ebola. WHO declined to fly her out of the country for treatment. *Washington Post*, September 15, 2014. https://www.washingtonpost.com/news/morning-mix/wp/2014/09/15/sierra-leone-loses-fourth-doctor-to-ebola/?utm_term=.4bd1f19a14e3.

Calkin S. Ebola claims life of nurse who worked with Will Pooley. *Nursing Times*, September 12, 2014. https://www.nursingtimes.net/clinical-archive/infection-control/will-pooleys-nurse-colleague-dies-from-ebola/5074762.article.

Evans DK, Goldstein M, Popova A. Health-care worker mortality and the legacy of the Ebola epidemic. *Lancet* 2015;3(8):e439–e440. doi:10.1016/S2214-109X(15)00065-0.

Fofana U, Giahyue JH. Health workers strike at major Ebola clinic in Sierra Leone. *Chicago Tribune*, August 30, 2014. http://www.chicagotribune.com/lifestyles/health/chi-health-workers-strike-ebola-clinic-20140830-story.html.

Green A. Remembering health workers who died from Ebola in 2014. *Lancet* 2014. doi:10.1016/S0140-6736(14)62417-X.

Hammer J. My nurses are dead and I don't know if I'm already infected. *Medium*, January 12, 2015. https://medium.com/matter/did-sierra-leones-hero-doctor-have-to-die-1c1de004.

McMahon SA, Ho LS, Brown H, Miller L, Ansumana R, and Kennedy CE. Healthcare providers on the frontlines: A qualitative investigation of the social and emotional impact of delivering health services during Sierra Leone's Ebola epidemic. *Health Policy Plan* 2016;31(9):1232–1239. doi:10.1093/heapol/czw055.

Matanock A, Arwady MA, Ayscue P et al. Ebola virus disease cases among health care workers not working in Ebola treatment inits—Liberia, June–August 2014. *MMWR*, November 14, 2014. https://www.cdc.gov/mmwr/preview/mmwrhtml/mm63e1114a3.htm.

Sierra Leone Telegraph. Sierra Leoneans in despair as Dr. Willoughby succumbs to Ebola. December 18, 2014. http://www.thesierraleonetelegraph.com/?p=8253.

Vogel G. Ebola's heavy toll on study authors. *Science*, August 28, 2014. http://www.sciencemag.org/news/2014/08/ebolas-heavy-toll-study-authors.

WHO. Ebola in Sierra Leone: A slow start to an outbreak that eventually outpaced all others. January 2015. http://www.who.int/csr/disease/ebola/one-year-report/sierra-leone/en/.

Sites of Treatment outside Africa
in the 2014–2016 Outbreak

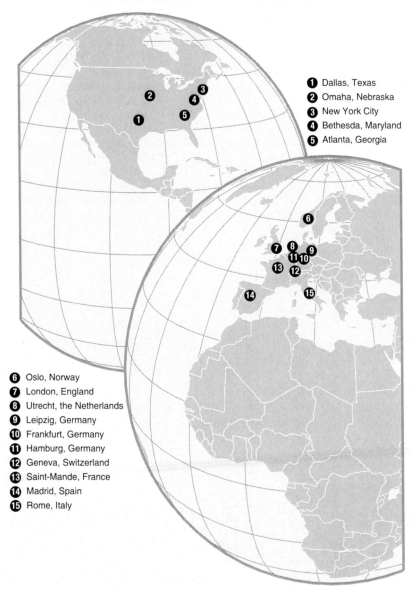

1 Dallas, Texas
2 Omaha, Nebraska
3 New York City
4 Bethesda, Maryland
5 Atlanta, Georgia

6 Oslo, Norway
7 London, England
8 Utrecht, the Netherlands
9 Leipzig, Germany
10 Frankfurt, Germany
11 Hamburg, Germany
12 Geneva, Switzerland
13 Saint-Mande, France
14 Madrid, Spain
15 Rome, Italy

MAP A.2   Ebola treatment sites outside of Africa. (Courtesy of Rod Eyer.)

PERSONS TREATED FOR EBOLA IN THE UNITED STATES AND EUROPE

A total of 26 people were publically reported to have been treated for Ebola outside of Africa during the 2014–2016 epidemics, based on reports from public agencies, hospitals, and news organizations. Most of them were U.S. or European citizens involved in relief efforts who were evacuated to their home country for treatment. Their mortality rate of 22% was far below the rate for the thousands of infected persons treated in West Africa. Of those treated in the United States or Europe, most arrived with their diagnosis already confirmed and all were treated in facilities that could provide intensive care and adequate fluid resuscitation. The sites of their treatment are shown in Map A.2.

COUNTRY OF NATIONALITY	ROLE IN OUTBREAK	PLACE DIAGNOSED	PLACE TREATED	OUTCOME
United States	Physician working with Samaritan's Purse aid group in Liberia	Monrovia, Liberia	Emory University Hospital, Atlanta, Georgia	Arrived in U.S. August 2, 2014 Released August 21, 2014
United States	Relief worker with Samaritan's Purse aid group in Liberia	Monrovia, Liberia	Emory University Hospital, Atlanta, Georgia	Arrived in U.S. August 5, 2014 Released August 19, 2014
Spain	Volunteer with Brothers Hospitallers of St. John of God in Liberia	Liberia	Hospital Carlos III, Madrid, Spain	Arrived in Spain August 6, 2014 Died August 12, 2014
United Kingdom	British nurse working in the relief efforts of Sierra Leone	Sierra Leone	The Royal Free Hospital, London	Arrived in UK August 24, 2014 Released September 3, 2014

(Continued)

COUNTRY OF NATIONALITY	ROLE IN OUTBREAK	PLACE DIAGNOSED	PLACE TREATED	OUTCOME
Senegal	Epidemiologist working for World Health Organization in Sierra Leone	Sierra Leone	University Medical Center Hamburg–Eppendorf, Hamburg, Germany	Arrived in Germany August 27, 2014 Released October 4, 2014
United States	Physician working with SIM international aid group	Liberia	Nebraska Medical Center, Omaha, Nebraska	Arrived September 5, 2014; Released September 25, 2014
United States	Physician volunteering with the World Health Organization	Kenema, Sierra Leone	Emory University Hospital in Atlanta, GA for treatment	Diagnosed September 6 Arrived at Emory September 9, 2014 Released October 19, 2014
France	Nurse with Doctors Without Borders (Médecins sans Frontières) in Liberia	Liberia	Bégin Military Hospital in Saint-Mandé near Paris, France	Arrived September 19, 2014 Release announced October 4, 2014
Liberia	Visitor to the United States. In Liberia he reportedly had driven a woman who later was found to have Ebola	Dallas, Texas	Texas Health Presbyterian Hospital, Dallas, Texas	Arrived in Dallas September 20, 2014 Diagnosed September 30, 2014 Died October 8, 2014
Spain	Physician, medical director at San Juan de Dios Hospital in Sierra Leone	Sierra Leone	The Hospital Carlos III in Madrid	Arrived in Spain September 22, 2014 Died September 25, 2014

(Continued)

COUNTRY OF NATIONALITY	ROLE IN OUTBREAK	PLACE DIAGNOSED	PLACE TREATED	OUTCOME
Uganda	Physician working for Emergency, an Italian aid group	Sierra Leone	University Hospital, Frankfurt, Germany	Arrived October 3, 2014 Released November 19, 2014
Spain	Nurse, cared for Spanish physician in Madrid	Alarcon, Spain	The Hospital Carlos III in Madrid	Diagnosed October 6, 2014 Released November 5, 2014
United States	U.S. freelance journalist working for NBC News in Liberia	Liberia	University of Nebraska Medical Center, Omaha, NE	Arrived October 6, 2014 in Omaha Released October 21, 2014
Norway	Physician working for Doctors Without Borders (Médecins sans Frontières)	Sierra Leone	Ullevål Hospital, Oslo, Norway	Evacuated October 6, 2014 Released October 20, 2014
Sudan	Laboratory technician at the UN Mission in Liberia, in charge of medical waste	Liberia	St. Georg Hospital in Leipzig, Germany	Became ill October 6, 2014; Arrived in Germany October 9, 2014 Died October 14, 2014
United States	Nurse cared for Liberian patient in Dallas	Dallas	National Institutes of Health Clinical Center, Bethesda, MD	Diagnosed October 11, 2014 Released October 24, 2014

(Continued)

COUNTRY OF NATIONALITY	ROLE IN OUTBREAK	PLACE DIAGNOSED	PLACE TREATED	OUTCOME
United States	Nurse cared for Liberian patient in Dallas	Dallas	Emory University Hospital, Atlanta, GA	Diagnosed October 14, 2014; Released October 28, 2014
United States	Physician, worked with Doctors Without Borders in Guinea	New York City	Bellevue Hospital Center, NYC	Arrived home to NYC October 16, 2014 Diagnosed October 23, 2014 Released November 11, 2014
Sierra Leone; U.S. permanent resident	Physician, chief medical officer of Kissy United Methodist Hospital in Freetown, Sierra Leone	Sierra Leone	Nebraska Medical Center, Omaha, NE	Diagnosed November 10, 2014 Arrived in Omaha November 15, 2014 Died November 17, 2014
Cuba	Physician working in relief efforts in Sierra Leone	Sierra Leone	Hôpitaux Universitaires de Genève (HUG), Geneva, Switzerland	Arrived in Geneva November 20, 2014 Released December 6, 2014
Not reported	UN worker	Sierra Leone	Bégin Military Hospital in Saint-Mandé near Paris, France	Released November 2014
Italy	Physician working for Emergency, an Italian aid group	Sierra Leone	Lazzaro Spallanzani Hospital, Rome, Italy	Arrived November 25, 2014 Released January 2, 2015

(Continued)

COUNTRY OF NATIONALITY	ROLE IN OUTBREAK	PLACE DIAGNOSED	PLACE TREATED	OUTCOME
Nigeria	Soldier served in UN Peacekeeping Force in Liberia	Liberia	University Medical Center Hospital, Utrecht, the Netherlands	Evacuated December 6, 2014 Declared cured December 19, 2014
United Kingdom	Aid worker at Ebola treatment center in Sierra Leone	Glasgow, Scotland	Royal Free Hospital, London	Diagnosed December 29, 2014 upon returning home; Released January 25, 2015 Rehospitalized October 2015, February and October 2016
United States	Clinician working for Partners in Health in Sierra Leone	Sierra Leone	NIH Clinical Center, Bethesda, MD	Admitted March 13, 2015 Released April 9, 2015
Italy	Nurse with Emergency, an Italian aid group	Sardinia, Italy	Spallanzani infectious disease clinic, Rome	Arrived home to Sardinia May 10, 2015 Hospitalized May 12, 2015 Released June 10, 2015

INFORMATION REGARDING OTHER SPECIAL PATHOGENS

A special pathogen is a highly infectious agent that produces severe disease in humans. Special pathogens include the viruses that cause several hemorrhagic fevers, such as Ebola virus disease (EVD), Lassa fever, and Hantavirus pulmonary syndrome, and other recently identified and emerging viral diseases, such as Nipah virus encephalitis. These viruses are RNA-coded (often negative-stranded or ambisense in coding strategy) and encased in a lipid envelope. All of the viruses are vector-borne zoonotic agents, meaning that under normal conditions, these viruses exist in animals. The majority is found in rodents, but some occur in other mammals or arthropods as well. All of these viruses are classified by the CDC as Biosafety Level 4 (BSL-4) pathogens and as such must be handled in special facilities designed to contain them safely. The *coronavirus* that causes SARS (severe acute respiratory syndrome) is also considered as a special pathogen.

Some important features of outbreaks of special pathogens include the following:

- Health care workers are usually unfamiliar with them, how they spread, how to diagnose them, and there is often limited local expertise.
- They are often accompanied with a quickly changing set of circumstances, confusing initial information, and high levels of fear and anxiety.
- Public health channels like the city and state departments of health and the U.S. Centers for Disease Control and Prevention are taxed to create a broad response.

They often require the rapid development of educational material for health care staff, patients, and communities. In addition, they require timely use of procedures to prevent spread of the pathogen within the health care facility and to protect employees, other patients, and visitors as the patient is treated. Outbreaks of special pathogens are occurring with some regularity. Over the past 20 years, the U.S. health care system has seen the arrival of Ebola, West Nile fever, SARS, MERS (Middle East respiratory syndrome), pandemic influenza, the anthrax attacks of 2001, and now Zika. We should anticipate that this pattern will continue.

How Should We Prepare for Future Outbreaks of Special Pathogens?

The appearance of special pathogens is a regular occurrence. Recent years have seen the unexpected spread of severe acute respiratory syndrome (SARS) and Middle East respiratory syndrome (MERS), both caused by novel respiratory viruses. In addition, mosquito-borne viruses, including chikungunya, West Nile virus, and Zika virus, have arrived in new areas of the globe, raising significant concerns. Concerns regarding pandemic influenza were sparked by the 2009 outbreak of a novel strain of that virus, one that had not been encountered for decades. Avian influenza, which carries a higher risk of death, remains an ongoing concern. Intentional release of infectious agents, as in the anthrax release in the United States shortly after the September 11, 2001 attacks, raised the ongoing specter of bioterrorism employing conventional or special pathogens. Finally, the ongoing reality transmission of foodborne pathogens, such as *Salmonella* or *Listeria*, in small and large outbreaks consumes substantial public health resources.

Preparing for the ongoing risks while remaining alert to unique, unexpected outbreaks of emerging or reemerging pathogens is a complex, formidable challenge. Lessons learned from many of the events listed above suggest the prudence of an all-hazards approach combining early detection, rapid contact tracing, and mitigation of the risk of further spread.

Index

Note: Page numbers followed by f and t refer to figures and tables, respectively.